trailblazing mars

UNIVERSITY PRESS OF FLORIDA

Florida A&M University, Tallahassee
Florida Atlantic University, Boca Raton
Florida Gulf Coast University, Ft. Myers
Florida International University, Miami
Florida State University, Tallahassee
New College of Florida, Sarasota
University of Central Florida, Orlando
University of Florida, Gainesville
University of North Florida, Jacksonville
University of South Florida, Tampa
University of West Florida, Pensacola

TRAILBLAZING mars

NASA'S NEXT GIANT LEAP

PAT DUGGINS

University Press of Florida

Gainesville · Tallahassee · Tampa · Boca Raton

Pensacola · Orlando · Miami · Jacksonville · Ft. Myers · Sarasota

15   14   13   12   11   10   6   5   4   3   2   1

Library of Congress Cataloging-in-Publication Data
Duggins, Pat.
Trailblazing Mars : NASA's next giant leap / Pat Duggins.
p. cm.
Includes bibliographical references and index.
ISBN 978-0-8130-3518-5 (alk. paper)
1. Mars (Planet)—Exploration. 2. Space flight to Mars.
3. Mars probes. 4. United States. National Aeronautics and
Space Administration—History. I. Title.
TL799.M3D84   2010
629.43'543—dc22   2010015139

The University Press of Florida is the scholarly publishing agency for the
State University System of Florida, comprising Florida A&M University,
Florida Atlantic University, Florida Gulf Coast University, Florida In-
ternational University, Florida State University, New College of Florida,
University of Central Florida, University of Florida, University of North
Florida, University of South Florida, and University of West Florida.

University Press of Florida
15 Northwest 15th Street
Gainesville, FL 32611-2079
http://www.upf.com

To Lucia—

For twenty years of ongoing patience while I chased rockets

contents

trailblazing mars

proLogue

A lot of kids dream of being astronauts. A lot of astronauts dream of going to Mars.

Long before the National Aeronautics and Space Administration (NASA) launched the last of its Space Shuttle missions to piece together the International Space Station, members of the astronaut corps had made Mars something of a hobby. "Mars is fun," says four-time Shuttle crew member Scott Horowitz. "To walk on a foreign planet, that's something I'll never get to do. But I'd like my kids to get to do that." How would astronauts get to the red planet? How would they survive there? How would they get back? These are just some of the questions Horowitz has thought a lot about on an unofficial basis. Just mention the subject and a twinkle appears in his eye. Horowitz isn't alone, and expectations about Mars have driven the debate as much as concrete science.

One pivotal obsession with Mars was due to a mistranslation. In the late nineteenth century, Bostonian businessman Percival Lowell was captivated by written accounts from Italian astronomer Giovanni Schiaparelli that spoke about "*canali*" on Mars. He meant "channels," but in Lowell's mind the phrase meant "canals," like the Suez Canal dug on purpose about the same time here on Earth. That inspired Lowell to write three books about the possibility of life on Mars and to build the Lowell Observatory in Flagstaff, Arizona. His passion about Mars also helped to fuel the popular belief that Earth's nearest neighbor was home to waterways, jungles, and perhaps inhabitants. The world of popular culture was also quick to catch the "Mars bug."

A 1919 first edition of Edgar Rice Burroughs's book *The Warlord of Mars* sits on my bookshelf at home. It was the third of his tales featuring gentleman adventurer John Carter on the red planet, known by its inhabitants as Barsoom. Burroughs created an exotic world where his fictional hero flew small, gas-filled airships called fliers to rescue the beautiful princess Dejah Thoris from the scheming villain Matai Shang. All the while, Carter's faithful ten-legged Martian hound named Woola fought at his side.

The image of the red planet took a more comic turn in 1948 when Warner Brothers introduced Marvin the Martian, the latest nemesis of Bugs Bunny. The popular cartoon character had a head like a black bowling ball and a costume including a Roman gladiator's helmet and oversize tennis shoes. In the debut episode, the "wascally wabbit" foiled Marvin's plans to blow up Earth with a bomb called the "Illudium Q-36 explosive space modulator." It seems our home world obstructed his telescope's view of Venus. "Where's the kaboom?" asked Marvin after Bugs stole the firecracker-shaped device. "There was supposed to be an Earth-shattering kaboom!" The cartoon alien would later cross swords with Daffy Duck in the cartoon "Duck Dodgers in the 24th and ½ century" in 1953.

That same year, Hollywood delivered a darker vision of life on Mars with a movie version of H. G. Wells's novel *War of the Worlds*, starring Gene Barry and Les Tremayne. Malevolent Martians using floating attack vehicles slashed their way through the world's major cities with heat-ray guns. Filmmakers used an electric guitar to create the pulsating sound effect of the aliens' weaponry.

Science fiction author Ray Bradbury took Burroughs's fanciful notions of adventures on Mars and placed them in the context of the Cold War with his book *The Martian Chronicles*, published in 1950. In it, Bradbury depicted efforts to colonize Mars as Earth decimated itself through nuclear war. He leans away from Wells's malevolent Martian invaders in *War of the Worlds*. Instead, the native population of Mars is destroyed by diseases brought from Earth. One of the colonists later stares down into the reflection of himself and his son and daughter in a pool of water. He declares prophetically, "There are the Martians," referring to themselves. Bradbury's book stitches together a collection of short stories, much in the same vein as his work *The Illustrated Man*, which came out the next year.

Mission patches created for NASA's two Mars Exploration Rovers. MER-A, later to be known as *Spirit*, was symbolized by the round patch featuring the Warner Brothers cartoon character "Marvin the Martian." MER-B, later called *Opportunity*, had the patch with Daffy Duck as "Duck Dodgers in the 24th and ½ Century." Courtesy of NASA.

In 1963, television viewers returned to the lighter side of Mars as they crowded in front of their TV sets to watch *My Favorite Martian*. "Uncle Martin," played by veteran Broadway actor Ray Walston, worked episode after episode to return to his home world after crash landing on Earth. Instead of the propeller-driven fliers envisioned by Burroughs or the impenetrable Mars vehicles in *War of the Worlds*, the silver spaceship people saw on TV was shaped like a duck's bill. While audiences chuckled their way through these adventures, events were under way that would drastically change how people on Earth viewed Mars.

In the early 1960s, the space race was under way. NASA struggled through the first launches of its one-man Mercury capsules that made astronauts John Glenn, Wally Schirra, and Gordon Cooper household names. While the public eagerly followed these events, scientists at the Jet Propulsion Laboratory in Pasadena, California, had trained their eyes on Mars. The *Mariner 4* spacecraft was launched toward the planet in 1964. Months later, it beamed back grainy television images that revealed no Marvin the Martian, no Uncle Martin, no Barsoom.

Still, NASA is making plans to visit there, and pop culture's fascination with Mars remains ever present. When the space agency launched the two Mars Exploration Rovers in 2003, Duck Dodgers and Marvin the Martian went along for the ride. The golf cart–size robots are now known as *Spirit* and *Opportunity*. Following their launch from Cape Canaveral, they had the less imaginative names of MER-A and MER-B. However, the Jet Propulsion Laboratory did anoint the vehicles with two embroidered mission patches to create a competition between the two teams of rover scientists on Earth. One patch featured Daffy Duck as "Duck Dodgers," and Marvin the Martian was on the other. Among things the rovers would photograph on Mars were the so-called Columbia hills. They were named for the seven astronauts who died when Space Shuttle *Columbia* broke apart and burned up in early 2003.

One year later, President George W. Bush ordered the end of the Shuttle program. The U.S. space effort was refocused to return people to the Moon and, perhaps, send astronauts to Mars. Venturing to the fourth planet in our solar system represents a chance for NASA to make up for budget cuts and questionable policy decisions that left the U.S. space program trapped in low Earth orbit since 1972. The expensive Apollo program was scrapped in favor of the Space Shuttle. Instead of a fully reusable two-stage space plane, budget cuts forced designers to attach the spacecraft to an expendable external fuel tank and two somewhat reusable solid-fuel booster rockets. A flaw in the rockets was blamed for the *Challenger* accident in 1986 that killed seven astronauts. A flaw in the external fuel tank caused the 2003 *Columbia* disaster, where a second crew of seven astronauts died. Along with the dangerous design, NASA was also criticized for being stuck in low Earth orbit, with no chance of pursuing more ambitious missions of exploration that the Shuttle couldn't do.

A mission to Mars could be NASA's new lease on life, but the American public may not be prepared for the risks. Landing humans on Mars is like nothing any space agency on Earth has yet attempted. The first man to walk on the Moon may have said it best. Apollo 11 astronaut Neil Armstrong was joined by crewmates Edwin "Buzz" Aldrin and Michael Collins at the Kennedy Space Center in July 1989. The event was to observe the twentieth anniversary of the first lunar landing. Armstrong thanked the thousands of people who worked for years so that he and his crewmates could go on what he called a "summer vacation." He wasn't far off the mark.

Even the hazardous first trips to the Moon were comparatively simple compared to the huge distances people will have to traverse to make it to Mars. Up to now, astronauts have blasted off on lots of "vacations," while visiting the red planet will be more like pioneering the Old West in the eighteenth and nineteenth centuries. Back then, people had to fight to get there and fight to survive there. The big difference is that the nation didn't have to watch the fate of the Donner Party on CNN.

Like those early pioneers, Mars crews will be largely isolated and cut off from help from civilization. The astronauts will have to repair their own damaged equipment, grow their own food, and deal with their own personal crises. Missteps along the way could result in tragedies that mission control in Houston won't be able to resolve. If Mars is next on NASA's agenda, the days of drinking Tang and waving to everyone on Earth during televised press conferences will be over. The era of true space travel will have begun, and vacation time will be over for NASA.

The notion of working alone and solving problems when it comes to Mars is familiar to the men who gave Earth its first disappointing views of the planet Mars. Their work in the early 1960s may have helped to pave the way in trailblazing Mars, NASA's next giant leap.

mariner sets sail

Nineteen sixty-two was the year Marilyn Monroe died. *West Side Story* won best picture at the Academy Awards, author Ken Kesey published *One Flew Over the Cuckoo's Nest*, and television viewers of NBC's *Tonight Show* were getting used to the program's new host, a little-known personality named Johnny Carson.[1]

Nineteen sixty-two was also the year of John Glenn, Scott Carpenter, and Wally Schirra. These three astronauts became the first Americans to rocket to orbit during NASA's Mercury program. But NASA was thinking about Mars as well.

While Americans responded to President Kennedy's challenge to put a man on the Moon, NASA would make its first step toward the red planet with unmanned probes. Future crewed missions, with astronauts facing challenges similar to those posed by the western frontier in the eighteenth and nineteenth centuries, would wait for a later date.

The process of launching robotic explorers to Mars would leave scientists at the space agency to cope with centuries of public fascination with Mars. Some vocal proponents hoped there might be a civilization on Earth's nearest neighbor waiting to welcome visitors with open arms. Building the first spacecraft to go to Mars was partly the job of a University of Iowa graduate student.

Stamatios Krimigis was known to his friends as Tom. He got the college assignment of his life while studying for his master's degree by working on radiation detectors. Krimigis was called to the office of one of his professors. Space exploration was still in its infancy, and the profession had few legends at that time. The short list of names included rocket pioneer Wernher von Braun. As a boy, his interest in space travel had been fueled by the science fiction writings of H. G. Wells and Edgar Rice Burroughs. The 1923 nonfiction work *Die Rakete zu den Planetenraumen* by Hermann Oberth asserted that space missions weren't simply the stuff of dreams. The young von Braun took that idea to heart and studied mathematics and later built rockets for Nazi Germany. Following World War II, he would accompany a small band of German rocket scientists to the United States to form the nucleus of NASA's efforts to send men to the Moon. Along with von Braun, the name of Krimigis's college mentor would be etched in space lore as well. He was James Van Allen.[2]

When the Soviet Union launched the basketball-size *Sputnik* in 1957, America was left unnerved as the battery-powered satellite sailed over New York and Washington, D.C., showering the United States with the rhythmic beeping of its radio transmitter. The White House wanted a quick response. Von Braun proposed sending up a thirty-pound U.S.-built artificial satellite on an army rocket. He also guaranteed he'd do it three months after the success of *Sputnik*. The concern had been raised about spaceflight being a stunt as opposed to an act of scientific merit. Van Allen created a radiation detector to ensure there was at least some scientific purpose to the mission of *Explorer 1*. The device discovered the belt of radiation, named after Van Allen, that encircles Earth.[3]

Now Mars beckoned.

Tom Krimigis knocked on the professor's door, not knowing what to expect. The room was cluttered with books, manuscripts, and photographs, and the legendary scientist sat at his desk. He wasted little time in stating what was on his mind.[4]

"How would you like to build an instrument for the first mission to Mars?" he asked. Specifically, Van Allen had something in mind called a trapped radiation detector. It would ride on an interplanetary spacecraft traveling to the red planet. During the lonely trip, the device would further study the Van Allen belt around Earth and see if something similar was going on around Mars. Krimigis responded that he didn't know how to build anything like that.

"That's alright," quipped Van Allen. "You can learn along the way."

Krimigis's opportunity followed a discussion between two NASA space centers on how best to explore the red planet. The fight was between the Goddard Space Flight Center in Maryland and the Jet

Propulsion Laboratory (JPL) at the California Institute of Technology in Pasadena. Goddard was the namesake of rocket scientist Robert Goddard, who conducted some of the first experiments on liquid-fueled rockets in the United States. The center that bears his name is perhaps best known for managing NASA's Hubble Space Telescope. JPL, on the other hand, is home to every major interplanetary mission NASA has attempted, including the two Voyager spacecraft to the outer planets and *Pioneer 10*, which became the first man-made object to leave the solar system. Back in the early 1960s, the space effort was new, and the pecking order at NASA was still being established.[5]

Venus or Bust

Goddard favored an ambitious mission to Mars featuring an orbiting spacecraft and a robotic lander. JPL suggested a less-demanding flight with a flyby trip using one of its Mariner spacecraft. The vehicle had already proven itself by traveling past Venus that year. A flyby of Mars meant the probe would coast past the planet, but not attempt to enter a stable orbit. There would certainly be no landing.[6]

Venus had long been one of the brightest objects visible from Earth, and also one of the most mysterious. Its thick, cloudy atmosphere shrouded the planet, leaving scientists to speculate about what surface features lay out of sight on the mustard-colored planet. During the 1962 Venus mission, *Mariners 1* and *2* each carried infrared and microwave instruments designed to try to pierce the cloud-tops. Powering these first two spacecraft was less of a concern since Venus is closer to the Sun than Earth. Two solar electricity panels would be attached to the probes to soak up sunlight. The data from Venus would be transmitted back to Earth on a circular high-gain antenna on a mast, resembling a big lollypop.[7] Four solar wings would be needed on the future Mariners going to Mars, where sunlight is much dimmer.

The flight of *Mariner 1* was also a test of the Atlas-Agena rocket that would carry the spacecraft. The idea of building two copies of a spacecraft wasn't to double the amount of data coming back to Earth. Instead, it was a way for NASA to hedge its bets because a launch-day disaster was a pretty good possibility. Rocket blastoffs from Cape Canaveral routinely ended in spectacular explosions, which provided a convenient, though expensive, fireworks show for residents as far east as Orlando. Losses mounted for nearly every type of rocket launched from the Cape, including one type called the Navajo, which locals nicknamed "the never-go." If one of JPL's Venus-bound Mariner vehicles didn't make it, its sister spacecraft might have a chance.

That turned out to be a smart move.

Mariner 1 blasted off on July 22, 1962, and it was a short trip that ended in failure. Every time a rocket takes off from either Cape Canaveral Air Force Station or the Kennedy Space Center, a range safety officer sits with his finger on the self-destruct button for the vehicle in case something goes wrong. Populated cities like Titusville and Cocoa Beach are perilously close to the launchpads at the Cape, so a wayward rocket could pose a genuine threat.

As *Mariner 1* blasted off aboard its Atlas-Agena, the rocket began to go the wrong way. The destruct button was pushed, and the rocket and *Mariner 1* were blown to bits. *Mariner 2*, waiting in the wings, would go next. The backup vehicle took off on August 27, 1962, and it survived the trip to space.[8]

The cruise to Venus turned out to be as much of a challenge as the blastoff. One solar panel failed, and the spacecraft began to overheat as it sailed closer and closer to Venus and the Sun. As *Mariner 2* glided past Venus at a distance of 21,000 miles, its instruments scanned the planet's cloud-tops and found them to be relatively cool compared to its broiling surface. Three weeks later, JPL lost radio contact with *Mariner 2*.[9]

Still, this was the first time a man-made spacecraft traveled to another planet, so the trip was considered a rousing success, and Mars was next. It also meant that Tom Krimigis's headaches were just beginning.

Aerospace contractor Martin Marietta, later to be known as Lockheed Martin, was hired to build the Mars spacecraft. It included an octagonal main body, which contained the radio transmitter and computers. Four solar panels formed an "X" across each vehicle. The one thing the main contractor wouldn't do is build the experiments that would go along for the ride. Those gadgets were considered the passengers to be added on by the scientists who envisioned them.[10] If the trapped radiation detectors Van Allen asked Krimigis to make for *Mariners 3 and 4* were going to fly, the grad student would have to do the work. "In those days, it was a do-it-yourself project," Krimigis recalled. "You did the electronics, you were the power supply person, you were the mechanical engineer who built it. Each investigator laid out the design of their own device, built the equipment by hand, and delivered it." He found a Chicago-based company that made the kinds of radiation detectors he needed, but there was a serious time crunch.[11] Krimigis began building his trapped particle device in January 1963, and the launch of *Mariner 3* was scheduled for November 1964. The mission would proceed whether or not the experiment was ready. "There's no way you could design, build, and approve an instrument for a NASA spacecraft in just sixteen months these days," says Krimigis. Back then, he was expected to do exactly that. The job meant he would see a lot of John Casani.

Casani joined JPL in 1962, and was named head of the design team for *Mariners 3* and 4. That meant when people like Tom Krimigis arrived with their science experiments, Casani was the traffic cop who would decide where the gadgets went on the spacecraft. "We developed an understanding," says Casani of the list of guest scientists on the mission. "We had to figure out where the instruments would go, how they would be bolted on, how much power they would need, and what wires went where." The result was a lot of cooks in the kitchen for a space mission that was more ambitious than NASA had ever attempted, even compared to the successful flight of *Mariner 2* to Venus.

That mission was a bit of a shortcut for JPL. The Venus vehicle was basically a copy of the Ranger spacecraft that had gone to the Moon.[12] Sending a probe to Mars would be a completely different job, and require a different machine. "It was pretty brand spanking new," says Casani, referring to the Mariner craft bound for Mars. "The structure was different, and we had a different destination. I mean going to the Moon only took a couple of days. Going to Mars would be a couple of months." Explaining the difficulties of a 1960s-era space mission required a 1960s explanation back then. When pressed for an answer by the public, Casani gave one inspired by the science fiction television show *The Outer Limits*. Each episode began with a mysterious narrator proclaiming, "We will control the vertical, we will control the horizontal." That referred to two little knobs on the front of the black-and-white television sets in many American homes. One controlled the vertical hold on the screen, and the other worked the horizontal. "We determined the complexity of Mariner 4 was like having 200 TV sets all lined up," recalled Casani. "The vertical and horizontal all had to work consistently all the time for the nine-month trip from the Earth to Mars. It was tough!"

All Tom Krimigis worried about was his little part of Mariners 3 and 4. His two radiation detectors resembled boxes with a "V"-shaped antenna like the "rabbit ears" you might see on an old television set. He delivered the prototype to JPL and breathed easy, for a little while anyway. Krimigis was back in Iowa, and sound asleep, when a phone call from California jarred him awake.

"Your detector's not working," said the voice on the other end of the line.

That prompted a flight back to Pasadena, so Krimigis could pull his device off the spacecraft to see what was wrong. "It turned out to be a loose connection," he says.

The Changing Face of the Space Program

NASA and the nation underwent great triumphs and tragedies while Mariners 3 and 4 were being readied for flight. Astronaut Gordon Cooper made the last flight of the Mercury program. He helped to pilot his Faith 7 space capsule during the perilous and fiery reentry into Earth's atmosphere. He would be the last American to fly in space alone. Astronaut Alan Shepard lobbied President John F. Kennedy for one extra flight, so he could gain experience in orbit before NASA moved on to manned missions to the Moon during Apollo. The White House said no, and the country moved on to building the two-man Gemini capsules, which would hone the skills of rendezvous and piloting that would be needed to put men on the lunar surface.[13]

Several gunshots in Dallas ended the life of NASA's greatest advocate at that time. The death of President Kennedy put Lyndon B. Johnson in the oval office, and some noticeable changes occurred at NASA. The Launch Operations Center along Florida's Atlantic coast was rechristened to be the John F. Kennedy Space Center. A much bigger surprise was coming for the residents of the nearby city of Cape Canaveral. The town gained world notoriety during the days of the Mercury missions

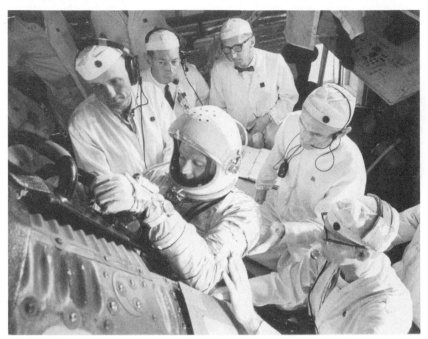

Astronaut Gordon Cooper squeezes inside the Mercury capsule dubbed *Faith 7*. This blastoff would draw the Mercury program to a close, and Cooper would be the last American to fly in space by himself. Courtesy of NASA.

because that's where reporters wrote their stories. There was no main media center near the launchpads where the press could work. That meant when journalists finished recording their stories for air on radio or television, these reports concluded with the reporter stating that he or she was broadcasting "from Cape Canaveral." Following the death of JFK, President Johnson arbitrarily renamed the town "Cape Kennedy," a decision that was reviled by those in local government as tinkering with Florida history. The change would stay in place until the conclusion of the Apollo program in the mid-1970s. At that point, residents of Cape Kennedy switched the name back to Canaveral.[14]

Despite the shake-up, engineers at JPL pressed on with preparations to launch *Mariner 3*. Liftoff day came on November 5, 1964. The spacecraft was perched on top of the now familiar Atlas-Agena rocket and encircled with a protective nosecone called a fairing. After liftoff, the fairing was designed to split open and fall away, leaving the probe exposed and ready to make the trip to Mars.

At least that's what was supposed to happen. The Atlas blasted off, but the newly redesigned fairing failed to open properly. The spacecraft remained trapped inside. Worse yet, its solar panels couldn't open to supply power. The death of *Mariner 3* would be the first in a series of high-profile failures of missions to Mars. At one point, only one out of every three flights to the red planet ended successfully. The 66 percent failure rate only served to emphasize the future dangers facing astronauts, assuming that NASA ever attempts a manned landing on Mars.[15]

Mariner 3 was lost, and once more JPL would depend on the backup craft to try it again. *Mariner 4* was fitted with an older and more reliable fairing, and the launch, on November 28, went ahead without incident. Now the waiting game began as the scientists and engineers at JPL sat patiently for *Mariner 4* to reach Mars. Krimigis filled some of that cruising time measuring the Van Allen belt and two solar flares with the trapped particle detector. He admits this was the part of the trip where

William Pickering, director of the Jet Propulsion Laboratory, delivers a scale model of NASA's *Mariner 2* spacecraft to President John F. Kennedy. This mission proceeded *Mariner 4*'s historic trip to Mars. Courtesy of NASA.

faith played a big role. "Every part on the spacecraft had to be tested to make sure it would work," says Krimigis. "But once it was on its way, the big worry was whether or not it was going to fall apart."

As technically advanced as the *Mariner 4* mission was at the time, there was one key element over which neither NASA nor JPL had control. No one was sure what part of Mars the spacecraft would sail over. Launch managers simply aimed for the planet and hoped for the best.

Perhaps the highest-profile experiment on *Mariner 4* involved the television camera, which was designed to take pictures of whatever part of Mars the spacecraft could see. But once the trip was under way, mission managers began to worry about what would happen if the camera broke down before the Mars encounter. That prompted some quick thinking, and another scientist at JPL would soon find himself in the middle of another historic moment on *Mariner 4*.

Arvydas Kliore came to the United States from Lithuania in 1948 to pursue a career in engineering. He earned his undergraduate degree from the University of Illinois in 1956, followed by his doctorate from the University of Michigan in 1962. After that, he received a phone call from a college chum who had moved to Southern California.[16]

"He talked about the great things that were doing at the Jet Propulsion Laboratory, and it sounded good," says Kliore. The notion of beaches and palm trees also seemed a lot better than brutal winters in Michigan, so Kliore decided to make the move to Pasadena and a job at JPL. However, his first assignment had nothing to do with the *Mariner 4* mission to Mars. Instead, he joined what the people at JPL called the "Future Projects Group."[17] That meant brainstorming about space trips, including a proposed mission where a spacecraft would hover at a spot between the Earth and the Moon called the libration point. That's where the gravity between the two bodies evens out, and the spaceship isn't pulled toward either one.

"They had me working on the mathematics on fuel consumption needed to keep the craft steady," says Kliore.

Then he got a phone call about some experiments he had done in college called radio occultation. JPL was thinking of using it as a backup in case the television camera on *Mariner 4* stopped working. The experiment would try to measure the composition of the Martian atmosphere by lining up Earth, *Mariner 4*, and the red planet. The probe's radio transmitter would be aimed at the very edge of the Martian horizon, with Earth as the target of the signal on the other side. The point was to shoot radio waves through the planet's atmosphere. Receiving stations on Earth would be waiting for the transmission to arrive so scientists could study what kind of interference was created by the air around Mars.[18] Ideally, the radio occultation experiment would reveal the density of the planet's atmosphere, which wasn't known at the time. The test might also confirm what gases were there. Astronomers using ground-based telescopes had studied the chemical gases in the Martian atmosphere. They concluded that a lot of it was carbon dioxide, but no one could prove it. The radio occultation experiment was thought to be able to provide the answer. But the catch was *Mariner 4* wasn't designed to do the test. If NASA approved the idea, the spacecraft would have to be reprogrammed out in space to perform the extra task, which had never been tried before.[19]

JPL asked Arv Kliore to pitch the idea to NASA.

"They liked previous presentations I gave," he recalled. Kliore also remembered a lot of butterflies in his stomach while preparing to sell the radio occultation experiment to members of the space agency. He took a plane to Washington, D.C., and geared up to face a conference room full of NASA personnel.

"I put on a full 'dog and pony' show for them," said Kliore. He remembered some sleepless nights at his hotel room as he worked the kinks out of his talk. On the day of his presentation, he fired up his viewgraph to show slides of the occultation proposal. That's when he noticed that, along with key NASA engineers, the associate administrator in charge of science for the space agency was in the audience as well. It was expected to be a tough crowd with the "big cheese" in attendance.

"It was pretty tense," Kliore recalled. He knew that NASA wouldn't just throw up its hands and approve the change immediately following his talk. The final word came after Kliore made a long flight back to Pasadena.

NASA eventually said yes.

Kliore's bosses at JPL apparently were so impressed with how he spoke before an audience that they planned to call on his talents once again. Crowds of reporters began to converge on Pasadena as *Mariner 4* drew closer to its target. Prior to the world's first mission to Mars, the attention of the press was focused on the original seven Mercury astronauts. In mid-1965, journalists were pounding on JPL's door. "For weeks prior to the encounter, newspapers were writing stories about the mission," Kliore recalled. "The [JPL] auditorium was filled with television cameras, movie cameras, and reporters. There were press conferences almost round the clock." And Kliore was one of the people who were asked to take questions and to deal with expectations, which were growing as fast as the crowds.

Dealing with the Press and "the Believers"

Prior to the 1965 arrival of *Mariner 4* to Mars, speculation ran rampant on whether life existed there. Astronomers had trained their telescopes at Mars for centuries, seeking clues to its surface. *Mariner 4* was the first opportunity to see the planet up close. Supporters of the idea of alien life thought their day was coming. Theories ranged from bacteria to plant life to full-blown civilizations waiting to wave "hello" to *Mariner 4*. A lot of this recent excitement was due to what appeared to be changes in the color of the surface of Mars from season to season as seen from Earth.[20]

In his 1962 book, *Mars: The Photographic Story*, astronomer Earl Slipher of the Lowell Observatory speculated that the changes had to be the result of something growing on the Martian surface. "Considering these points, which are the result of many kinds of observations," he wrote, "some kind of vegetation [on Mars] exists."[21]

The same changes in the appearance of Mars prompted a similar article by Frank Salisbury in *Science* magazine in 1962. "Of all of the proposals for the observed Martian phenomena," he wrote, "the idea of life on Mars seems the most tenable." The major newspapers seemed to take their cue from the speculators, saying that the existence of life on Mars would be proven or debunked by *Mariner 4*.[22] This left one group of people at JPL on the "hot seat." They were the members of the Mariner imaging team, and Bruce Murray was among them.

Murray was a veteran of the U.S. Air Force who had joined JPL in 1960. The use of television to take the first pictures of Mars was a technological leap at the time. It was also the first time digital TV had been used from a space probe. The images from *Mariner 4* would come back

NASA's *Mariner 4* spacecraft was the first manmade object to fly past Mars. The vehicle carried a digital television camera and transmitter as well as experiments to study the atmosphere and magnetic field around the red planet. Photo courtesy of NASA.

to Earth at an agonizingly slow eight bits per second. One complete shot would take up to eight hours to transmit. "It was like watching grass grow," recalled Murray. "The challenge was getting the equipment to work reliably far from Earth. The signal was really weak." Complicating the effort to capture some interesting views of Mars was the presence of "the believers." These were individuals who expected evidence of some kind of life on Mars, whether it was simple lichens or the more elaborate civilizations envisioned by Edgar Rice Burroughs and H. G. Wells. Murray considered himself a pretty skeptical scientist, but he admits growing up on fictional accounts of Mars featuring plenty of alien characters. "Mars had this special value," says Murray. "Writers had created this world, and there was a really powerful advocacy. Universities like Cal Tech were pretty friendly with this kind of culture." Still, Murray thought Mars was likely to be an arid and unfriendly place. The notion that *Mariner 4* would find anything alive, he thought, was pretty shaky.

Learning the truth, one way or the other, would require patience. *Mariner 4* passed within 6,000 miles of Mars on July 14, 1965, eight months after its liftoff from Cape Canaveral. The first target of opportunity was the plain known as Amazonis Planitia, just west of the Arizona-size volcano called Olympus Mons. *Mariner 4* then cut a swath

This photo was taken by *Mariner 4* during its brief flyby of Mars in 1965. Instead of civilizations and lush jungles, envisioned by Edgar Rice Burroughs, the robotic probe discovered Mars was barren like Earth's moon. Photo courtesy of NASA.

across Mars amounting to maybe 1 percent of its surface. The first shot would arrive after *Mariner 4* had passed the planet and the pictures had been stored in the spacecraft's onboard recorder.[23]

Arv Kliore was in the mission operations center at JPL as the first grainy image of Mars slowly appeared, one line at a time, on a large television screen. It was the planet's horizon.

He and the assembled scientists and reporters all asked, "Is that all?"

It took months of work to assemble *Mariner 4* and design its instruments. Engineers faced the loss of *Mariner 3* and the hazards of launch day for the backup craft. Then they had to hold off the onslaught of the media and the expectations of "the believers" during the trip to Mars.

For that effort, the planet looked boring.

"There were none of the rift valleys like Valles Marineris, or the majesty of Olympus Mons," recalled Kliore. "There were just craters. It looked like the Moon."

"It was pretty cruddy," stated Bruce Murray from the imaging team.

Worse yet, information from the instruments on *Mariner 4* indicated that Mars was dead. The trapped radiation experiment built by Tom Krimigis showed no radiation around the planet. The Van Allen belt protects Earth from ultraviolet rays streaming from the Sun. No shield like that encircled Mars, so the harmful rays rained down on the surface relentlessly. Arv Kliore's radio occultation experiment shot radio waves through the Martian atmosphere and found it was mostly carbon dioxide, and a thousand times thinner than the air on Earth. Mars was not the exotic paradise envisioned by Edgar Rice Burroughs, and the news reports streaming from JPL reflected it.[24]

The *New York Times* described the twenty-one pictures from *Mariner 4* as a "harmful and perhaps fatal blow to the possibility of current or past life on Mars." The headline in the *Times* on July 30 read, "Mariner's Close-Range Photos Indicate That Mars, Like the Moon, Is Lifeless Crater-Pocked Planet."[25]

Ouch.

Still, the imaging team at JPL worked to make the most of what *Mariner 4* had delivered. One reason was that JPL director William Pickering was intently watching the progress of the flight. He had made the cover of *Time* magazine in March 1963 with the successful flyby of *Mariner 2* to the planet Venus. He was depicted gazing upward at the planet while the face of the Greek goddess of love, for whom the planet is named, gazed lovingly back.[26]

The television pictures of Mars from *Mariner 4* were grainy at best, and in black-and-white. Some way had to be found to depict the planet in color. Scientists knew shades of red and tan were present on the planet from observations taken with telescopes on Earth. So something had to be done to match expectations with what was coming back from Mariner.

JPL used crayons.

Technicians took the Mariner photos and filled in the colors, mostly with tan and brown crayons, and then presented them to Pickering.

Time magazine from July 23, 1965. The cover featured William Pickering, director of the Jet Propulsion Laboratory, following the successful flyby of *Mariner 4* past Mars. Courtesy of Time Warner.

"He was happy with that," remembered Arv Kliore. "In fact, one of the shots was framed and hung outside his door for months."

Better yet, Pickering made the cover of *Time* magazine again, this time in the July 1965 issue, which coincided with *Mariner 4*'s flyby of Mars.[27] Still, the photographic results from the mission were a big letdown. "It was a terrific disappointment for the scientists and the public," says Tom Krimigis. "Now, Mars wasn't like in the H. G. Wells novel, no people, no canals."

The big concern among the members of Mariner's mission management team was the reaction in Congress. The absence of mind-boggling results might dampen interest in funding future missions to Mars, and perhaps the other planets. The worry turned out to be largely unfounded. Six more Mariner spacecraft would be built and launched to study Venus and Mercury, and to return to study the cratered surface of Mars.

There was, however, another controversy connected to the mission of *Mariner 4*. When the U.S.-built spacecraft made its journey to the planet Mars, it wasn't alone.

the space race
to mars

By the time *Mariner 4* was being readied for liftoff, the Soviet Union had already established itself as a leader in space. Cosmonaut Yuri Gagarin became the first human to orbit Earth in 1961. Gherman Titov further widened Moscow's lead with a daylong spaceflight later that same year. Astronaut John Glenn wouldn't become the first American to circle the globe until 1962, and Gordon Cooper had to wait until 1963 to duplicate Titov's marathon mission.[1]

Soviet ambitions didn't end there. The Kremlin wanted to conquer Mars, but Moscow would face many of the same setbacks and disappointments that dogged NASA's efforts. When Lewis and Clark and Zebulon Pike led their early expeditions to explore the American West, Spanish settlements and missions already dotted what would become California. In this same vein, the Soviet Union had its eye on Mars, but that ambition would come at a cost.

Unmanned Russian probes were launched toward the red planet starting in 1960, the year before Gagarin's history-making trip to space aboard the *Vostok-1* spacecraft. This includes one blastoff on October 24, 1962, during the height of the Cuban missile crisis. While President John F. Kennedy and Soviet premier Nikita Khrushchev sparred over nuclear weapons on the Communist island, a Soviet spacecraft bound for Mars exploded over the North Pole. The spectacular loss of the vehicle, known simply as 2MV-4 Number 3, prompted momentary concern over a possible Soviet nuclear attack. The debris from the blast came within view of the Pentagon's Ballistic Missile Early Warning System, or BMEWS, in Alaska.[2] Since something akin to World War III did not occur, observers suggest the failed Russian launch was quickly judged as nonhostile.

Mariner 4 Has Company

On November 30, 1964, two days after NASA launched its first space-craft to successfully reach Mars, a Russian rocket blasted off. The Soviet space program was typically secretive at that time, but the fact that *Mariner 4* had a competitor on its way to Mars appeared plain. Radio signals from Russian ground controllers to the unmanned spacecraft, called *Zond-2*, were picked up by the Jodrell Bank radio telescope in Manchester, England, and even by NASA's own fledgling spacecraft-tracking network.[3]

Mariner 4's mission to photograph Mars and conduct experiments to study its atmosphere and detect possible radiation belts was well documented. *Zond-2*'s objectives were, officially, more cloudy. The spacecraft was believed to carry a camera system, like *Mariner 4*'s, but also a separate landing vehicle intended to parachute to the surface of the red planet.[4] If successful, Russia would have upstaged NASA once again. In 1966, the United States built *Surveyor 1*, which would be the first NASA craft to soft land on the Moon. Moscow's alleged goal was to put something on Mars the year earlier.

Official accounts disagree over whether *Zond-2* actually carried a landing probe. There's speculation that the spacecraft's slow speed to-ward Mars was meant to put the vehicle on a closer path to the planet compared to *Mariner 4*, which some contend could have allowed *Zond-2* to drop a probe on a parachute. Even if this turned out to be true, the probe may not have survived the trip down to the surface of Mars. Russia and NASA operated under the belief that the air surrounding the planet was much thicker than was the case. *Mariner 4* later revealed Mars's atmosphere to be much thinner. Assuming *Zond-2* carried a lander equipped with a parachute, students of the Soviet space program believe it would have likely crashed on the surface. That's because mission designers, under the belief that Mars had a relatively dense atmosphere, would have used a chute that would have been ineffective in the planet's thin layer of air.[5]

Zond-2 is believed to be a Soviet space-craft launched at nearly the same time as *Mariner 4* on a mission to Mars. There's speculation that the vehicle carried a landing probe designed to descend by parachute to the planet's surface. Photo courtesy of NASA.

Photographs of *Zond-2* released by Russia reveal a clumsier-looking spacecraft compared to the graceful lines of *Mariner 4*. The body of the Soviet vehicle resembled a refrigerator compressor with a round wire mesh antenna to send and receive commands and data.[6] However, observers of the Soviet space program who pore over the snippets of information that leaked out over the years, before and after the fall of the Iron Curtain, credit Russia for some sophisticated ideas.

Mariner 4 took photos with a television camera and beamed the pictures back with a digital signal. *Zond-2* was reportedly equipped with a traditional film camera that was designed to snap shots of Mars during its flyby mission. The film would then be processed inside an onboard darkroom, scanned, and then beamed back to Earth.[7]

This brand of detective work regarding the "cloak and dagger" world of the Soviet space program appears to be what really fires up observers of those days. They point out even the name *Zond-2* is an example of how secretive Moscow was at that time. "Zond," they contend, is from the French word "*sonde*," simply meaning probe. Most of the thousands of spacecraft launched by the Russians carried names like "Zond" or "Cosmos."[8] By comparison, few missions were dubbed "Mars" or "Venera," which is Russian for Venus. Observers note that such an obvious name would have tipped off the world as to where that spacecraft was going. Moscow reportedly preferred to keep things close to the vest. Take *Luna-1*, for example. That spacecraft was hailed by the Soviets to be the first manmade object to go into orbit around the Sun. Later documents revealed that the original intention was for the vehicle to hit the Moon; it clearly missed. Followers of the Soviet space program note how the objectives of the *Luna-1* flight were adjusted to turn a failure into another "first."[9]

The fate of *Zond-2* appears to be similar to that of *Luna-1*. Instead of becoming a much-anticipated "feather" in Moscow's cap, the probe vanished on May 18, 1965. The loss of the vehicle was blamed on a solar electricity panel that failed to open as designed. As the spacecraft moved further toward Mars and away from the Sun, its power levels dropped, and the vehicle died. Jokes were made around the Jet Propulsion Laboratory that *Zond-2* had been gobbled up by a "great galactic ghoul."[10] The loss of the vehicle was one of many failed Mars missions by NASA and the Russians. Even so, while NASA was fighting the Soviet Union in the race to the Moon, plans were under way to send Apollo-era astronauts to Mars.

reheating
the Leftovers
of apollo

In the 1960s, astronauts like the "Mercury Seven" were the public face of the space race. America's first space pioneers, with their buzz haircuts and skinny neckties, smiled from newspaper photos and magazine covers. Schoolkids at that time were swept up as part of the so-called Tang dynasty, named for the orange-flavored drink mix popularized by NASA. Television viewers couldn't escape the space race, either. Producers of the *Gilligan's Island* show wrote an episode where two wayward Russian cosmonauts crash land in their capsule, only to match wits with the castaways, including the Skipper, the Professor, and, of course, Gilligan.[1]

During this time, NASA was asking questions about sending people to Mars. A future trip to the red planet would include dangers and challenges similar to those faced on the western frontier in America in the eighteenth and nineteenth centuries. However, the United States wouldn't have to wait to deal with the pain of a fatal accident related to space travel. Astronauts Gus Grissom, Ed White, and Roger Chaffee died during a fire inside their space capsule in 1967, without leaving the ground. They were performing a launchpad test of the first Apollo craft to fly when the blaze broke out. NASA technicians raced to the scene of the accident in futility, only to find the astronauts dead. More disquieting was that the solid rocket booster on top of the capsule, which was supposed to pull the vehicle away in case of an emergency, was fully loaded with propellent. Engineers at the scene of the tragedy later

This is the damaged Apollo 1 spacecraft where astronauts Gus Grissom, Ed White, and Roger Chaffee died in a fire during a routine launchpad test. The accident put the Apollo Moon program on hold for months until the fire hazard was minimized. The capsule, however, still had to utilize the pure oxygen atmosphere, which fed the fire that killed the Apollo 1 crew. Courtesy of NASA.

The crew of Apollo 7, with astronaut Walter Cunningham at the far right. The mission put the Apollo mother ship through its paces, and earned NASA an Emmy award for TV telecasts from Earth orbit. Photo courtesy of NASA.

considered themselves lucky that they weren't barbequed by the rocket had it ignited due to the fire inside the Apollo capsule.

The deadly consequence of America's space ambitions could be repeated as humans are sent on longer and more complicated trips to the red planet. As tough as sending people to the lunar surface would be, it was a trip of only a few days. Going to Mars would require more money and months of constant danger compared to the mammoth task of Apollo. The American people had been sold on going to the Moon to defeat the Soviet Union. Going to Mars might have sailed along on Apollo's coattails if pitched the right way.

Preliminary studies to send people to the red planet were no secret at NASA or to the astronauts of Apollo, including Walt Cunningham. He and his crewmates, Donn Eisele and Mercury veteran Wally Schirra, were the backup crew to the ill-fated mission of Apollo 1 where Grissom, Chaffee, and White had been killed.

Cunningham and his crewmates would try to get the Moon effort back on track by blasting off in a redesigned Apollo capsule in 1968. The astronauts put the new vehicle through its paces, and even won a special Emmy award from the television industry for the first "live" TV broadcasts from space.[2]

After the success of Apollo 7, three more astronauts would circle the Moon during the Apollo 8 mission. Then, another crew would test the spidery lunar module that would eventually carry people to the lunar surface. During all of this work, the possibility of a Mars mission was lurking in the background. "We might have made it," recalled Cunningham. "There wasn't a man in the astronaut office during the Apollo days who didn't think we'd be on Mars by the turn of the century." NASA began dropping gentle hints in the press about sending people to the red planet shortly before Neil Armstrong and Edwin "Buzz" Aldrin touched down on the lunar surface. The agency also commissioned studies from contractors like Lockheed and TRW to examine what it would take to send expeditions of perhaps six astronauts on a Mars mission, by building on the infrastructure of Apollo. Internal reports, like one written by NASA's Marshall Space Flight Center in Alabama in 1965, proposed using Apollo rockets and capsules to mount a manned flyby mission of both Venus and Mars as soon as 1979.[3] That effort, the authors believed, would pave the way for a future manned landing on the red planet. While astronauts like Walt Cunningham were confident it would happen, others were less so.

Alan Bean was hired by NASA as part of the astronaut class of 1963. The space agency picks its crew members in groups, or classes. The first class was made up of the legendary astronauts of Project Mercury: Alan Shepard, Gus Grissom, John Glenn, Wally Schirra, Scott Carpenter, Gordon Cooper, and Deke Slayton.

Members of Bean's 1963 group gathered in gray and black business suits for their group photo. Bean sat in the middle of the front row, holding a small, white scale model of the Mercury space capsule that made NASA's first trips to orbit. The group included its own bits of history and tragedy. Sitting among the candidates were Michael Collins and Buzz Aldrin, who would take part in the first Moon landing during the Apollo 11 mission. The man seated on Bean's right was Charles Bassett, who would be killed in a training jet accident. Also in that row was Roger Chaffee, who would die in 1967 during the Apollo 1 launchpad fire.[4]

Astronaut Alan Bean works at the Ocean of Storms on the surface of the Moon during Apollo 12. Bean joined Pete Conrad on NASA's second manned lunar landing after Apollo 11. Courtesy of NASA.

Bean visited the lunar surface during the mission that came after the trip by Aldrin and Armstrong. That was Apollo 12 in 1969. Later, Bean went on to command the second manned flight of America's first space station, known as Skylab. He had heard the rumblings of a possible Mars flight as a follow-up to the lunar program in the 1960s.[5] He says that back then, the technology of the time wasn't up to the task that Mars planners had in mind. "You didn't have the supplies," Bean recalled. "You didn't have the propulsion." The Saturn V rockets that NASA used to send people to the Moon were a technological wonder of the world. Each was nearly 400 feet tall and so complicated that different contractors were hired to build the three sections, or stages. The rocket would blast off in one big piece. The first stage would drop away when its fuel ran out. Then the second stage could ignite. The final, or third, stage would finish the job of pushing the Apollo spacecraft to the Moon.

Boeing built the first stage, which carried the six-million-pound vehicle off the launchpad with over seven million pounds of rocket thrust. The second stage, built by North American Aviation, finished the trip to Earth orbit with its own engines generating a further one million pounds of thrust. The third stage, called the S-IV-B, was designed and constructed by Douglas Aircraft. It would send astronauts from Earth to the Moon with one engine capable of 200,000 pounds of jet power. The Saturn V was the biggest rocket NASA ever built.[6]

But as remarkable as this spacecraft was, it took every scrap of thrust it could muster just to go to the Moon. That prompted astronaut Alan Bean's skepticism about using Apollo to go to Mars. "It's like having a car with a twenty-five-gallon gas tank," says Bean, "and somebody says, 'Let's go from Houston to Los Angeles.' The problem is you run out of gas by El Paso. Great idea, but it won't work." However, one group of scientists thought they could make an Apollo-style Mars mission possible. Their strategy was to put a different kind of engine under the hood.

Project NERVA

NASA was created on October 1, 1958. The agency hired Harold Finger the same day. Within two years, he was put in charge of the office to develop nuclear-powered rockets. The civilian space agency was building on work done in 1955 by the U.S. Air Force and the Atomic Energy Commission (AEC). These entities teamed up with the Los Alamos National Laboratory, which had created the atomic bomb. The Lawrence Livermore Lab, which specialized in nuclear military weapons, was also involved. The goal of this effort was to create atomic engines for intercontinental ballistic missiles, but the program was killed in 1957, one year before NASA was born. Instead of weaponry, the U.S. space program was looking for ways to propel astronauts and space probes to the outer planets.[7] For Harold Finger, that meant one thing. "We

President John F. Kennedy addresses Congress about the future of the U.S. space program, including a challenge to create nuclear propulsion to carry astronauts to Mars and, perhaps, farther into the solar system. Photo courtesy of NASA.

did think in terms of Mars," says Finger. "I certainly did, and our whole group did."

Finger began his engineering career by working for the Lewis Research Center in Cleveland, Ohio, in 1944. The facility later became affiliated with NASA. Finger specialized in experimenting with superchargers from Japanese and German aircraft captured during World War II. He moved on to jet engines, and then came the opportunity to study nuclear power. Eventually, that led to NASA and a mandate from President John Kennedy.[8]

JFK's speech at Rice University in Texas in 1962 laid out his goal of sending an astronaut to the Moon by 1970 and returning him safely to Earth. An earlier address before Congress in May 1961 included much of the same rhetoric regarding his lunar program. But the president also recognized the limitations of rockets using liquid or solid fuel. Something else was needed. In his address, Kennedy also called for nuclear rocket propulsion, and he was prepared to spend $30 million to get the

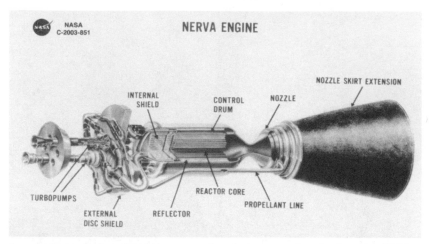

NASA
C-2003-851

NERVA ENGINE

NOZZLE SKIRT EXTENSION

INTERNAL
SHIELD

CONTROL
DRUM

NOZZLE

TURBOPUMPS

REACTOR CORE

PROPELLANT LINE

EXTERNAL
DISC SHIELD

REFLECTOR

An artist's conception of NASA's proposed nuclear-powered engine in project NERVA. The hydrogen-fueled motor was intended to propel an Apollo-derived spacecraft on a trip to Mars, either a flyby or landing trip. Courtesy of NASA.

program going. "This gives promise of some day providing a means for even more exciting and ambitious exploration of space," said Kennedy, "perhaps beyond the Moon, perhaps to the very end of the solar system itself."[9]

Finger had worked on earlier versions of nuclear propulsion called Project Rover, and a reactor called KIWI. "We called it that because it never flew," he recalled with a chuckle. Even though this kind of rocket used atomic power, it was simpler than the chemically fueled engines NASA planned to use on Apollo. Each of the stages of a Saturn V Moon rocket had two fuel tanks inside. One tank was for the propellent, and the other contained an oxidizer. The two would mix in a combustion chamber to create an explosive reaction. That was the thrust needed for liftoff.[10]

Nuclear rockets did away with that. "It's very straightforward," says Finger. "You have a nuclear reactor and hydrogen for the fuel." The reactor heats up the hydrogen, causing it to ignite and create thrust, which thunders from the nozzle at the rear of the engine. It closely resembled the motors in use for Apollo. Supporters of atomic rockets say they would be faster than the Saturn V rockets that went to the Moon and weigh only half as much. Less weight meant less fuel burned.[11]

NASA eventually teamed up with the AEC to try to create a program called Nuclear Engine for Rocket Vehicle Application, or NERVA for short. Ground-based tests using prototype motors were conducted first, with an eye on flight tests in space coming later. That early work led to a test engine in 1967 that ran successfully for over an hour at 1,100 megawatts of power. "That would have given you quite a lead on a trip to Mars," says Finger.

While the effort went on to build nuclear propulsion for a proposed Mars mission, a separate project to capitalize on the Apollo Moon effort was under way.

The Job No Astronaut Wanted

For the men who fought for the chance to walk on the Moon, being assigned to the Apollo Applications Program (AAP) was like wearing a big red button that read "no flight assignment."

Astronaut Alan Bean's first job was at AAP. "It wasn't a choice assignment," Bean admits. Ninety-five percent of the manpower at NASA was assigned to putting a man on the Moon before 1970. Agency managers said the remaining effort could be spent on AAP, so long as it didn't get in the way of the first Moon landing.

That meant astronauts like Bean wanted to work on one of two spacecraft, either Gemini or Apollo. The two-man Gemini vehicles sharpened NASA's ability to change orbits, rendezvous, dock, and do the other things that were needed to reach the Moon. Every astronaut who rode on a Gemini craft would later be assigned to an Apollo capsule, which would finish the job of going to the Moon. Al Bean wasn't assigned to work on either craft. That left him to help reheat the leftovers of Apollo.[12]

While NASA contractors hashed over the possibilities of sending people to Mars, the men in the AAP office found there was less spare hardware with which to work. That changed as the Apollo 11 Moon landing neared reality, because Congress was reducing NASA's budget, which refocused AAP to a program that relied on leftover parts from a shrinking Moon program. Job one for AAP wasn't going to Mars, but rather building America's first space station.[13]

There was a specific number of Saturn rockets built to carry astronauts to the Moon. Likewise, there was a limited inventory of the crew capsules, such as the gumdrop-shaped command modules where Apollo crews lived and worked on their journey to and from the Moon, and the buglike lunar modules that settled on the Moon's gray, dusty surface. Some of the spacecraft and rockets were used for test launches. Others were used on actual missions. AAP had to work with whatever was left. The list of Moon missions began with Apollo 7, which tested the command and service modules the astronauts would use to fly to the Moon. NASA wanted to fly missions all the way to Apollo 20, which would have carried astronauts Pete Conrad, Jack Lousma, and Paul Weitz to Tycho Crater on the Moon.[14]

The designers in AAP were working on a plan to hollow out the empty third stage of a Saturn Moon rocket and make it into an orbital workshop. Astronauts would then use the facility during longer periods in space. Since the third stage had once carried a full load of fuel, this plan was called the "wet tank" approach.[15]

The Saturn S-IV-B, also known as the third stage, was sixty feet long and twenty feet wide. To the engineers of AAP, it had "space station" written all over it. "We were thinking," says Bean, "why not launch it [the Saturn rocket], turn the command module around, dock it to the fuel tank, open the hatch, and float inside?"

Then reality set in.

Mission planners found the "wet tank" plan would require a lot of work in space. Astronauts would have to carry up oxygen tanks, equipment, as well as an airlock so they could float in and out while wearing heavy spacesuits to perform spacewalks. A planned astronomical observatory called the Apollo Telescope Mount (ATM) would be launched at a later date and snapped on the outpost. What had started off as a nifty notion to build a quick and dirty space station had turned into a technological tangle that even NASA couldn't unravel. "It was like saying 'I'm going to take my garage and turn it into a five-story office building,'" recalled Bean. "When you try to do it, you find you have to tear it down, build a new foundation, and bring in trucks." Again, it may have been a great idea, but NASA couldn't make it work. Fortunately for Bean, AAP wouldn't be his problem for much longer. Astronauts John

Young and Michael Collins were assigned to fly the two-man Gemini 10 mission in 1966. Al Bean was named backup commander, which started him on the path toward his walk on the Moon during the Apollo 12 mission. His responsibilities at AAP were passed along to other men, including astronaut Walt Cunningham, whom Bean credits with key decisions that led to the creation of Skylab.[16]

Cunningham, Bean, and all the other astronauts looked to one person for their chances to go into space. That was Deke Slayton, who was picked to fly on Project Mercury, but was grounded due to a heart murmur. He took the job of director of Flight Crew Operations at NASA, which meant he named the crew members for Gemini and Apollo flights. Cunningham had been assigned to the 1968 Apollo 7 mission along with Wally Schirra and Donn Eisele.[17] After he and his crewmates returned successfully to Earth, his next job wasn't a lunar flight. He was sent to AAP. "I got the assignment the week we got back," recalled Cunningham. "I think that was Deke's plan all along. He just didn't want to tell me before the [Apollo 7] flight." That meant Cunningham was out of the rotation for a Moon mission. He never flew in space again. However, budget cuts surrounding the successful Apollo 11 mission gave Cunningham and his AAP colleagues extra hardware to work with. Fewer dollars from Washington prompted the cancellation of the last three lunar landings. The result was a complicated shuffle of personnel and spacecraft as astronauts scrambled for the dwindling number of seats to the Moon.

Apollo 20 was the first to be canceled. Astronaut Pete Conrad would have led the three-member crew with Jack Lousma as lunar module pilot. These two men would have walked on the Moon while Paul Weitz orbited the Moon in the command module mother ship. All three men would find work orbiting Earth aboard Skylab, which Walt Cunningham was laboring to create.

After Apollo 20 was canceled in January 1970, the flights of Apollo 18 and Apollo 19 met similar fates. That cost astronauts Dick Gordon and Vance Brand their chance at a Moon flight, although their crewmate Harrison Schmitt was moved to Apollo 17, which did go. The end of Apollo 19 meant similar disappointments for Fred Haise, Bill Pogue, and Jerry Carr. Haise narrowly escaped death on Apollo 13 after the spacecraft was left crippled by an explosion halfway to the Moon. He never flew again. Like the crew of Apollo 20, Pogue and Carr were later named to a Skylab flight.[18]

Along with shuffling people, Walt Cunningham and his colleagues at AAP had new hardware to utilize. "We had long since decided that we wanted an orbital workshop," said Cunningham before his arrival at AAP. "It's just that we wanted the wrong workshop." He supported abandoning the plan to launch a Saturn V fuel tank and then convert it into an orbiting lab with the "wet tank" approach. Cunningham and a growing number of supporters wanted to claim the Saturn V rocket from a canceled Apollo mission. Engineers would turn the third stage into a space station while it was still on the ground. The interior would be divided into a working and living area. The "work" half had two airlocks and space for experiments. The "living" section had a dining room, sleeping bags, and a shower. The floor between the two areas was a metal grid with holes designed to fit the cleats on the astronauts' shoes. This was so the crew members could lock themselves in one place instead of floating around in the lack of gravity. The ATM would be installed on one end of the station near the spot where crew capsules would dock to bring in fresh crews and supplies. That part was called the multiple docking adapter. Power would be provided by X-shaped solar electricity panels near the ATM and two rectangular panels along the side of the lab.[19]

The facility, now known as Skylab, would be launched into orbit in one piece ready to go. "We kept on getting converts," says Cunningham, "until everybody thought it was the only way to go." The idea received final approval in 1970, and its first mission was launched in 1973. More leftovers from NASA's canceled Apollo flights would be used to transport the three crews to and from the orbiting laboratory. The capsule from Apollo 18 was selected to launch the first Skylab crew to orbit. Their mission, called Skylab 2, would last twenty-eight days. That was double the previous U.S. endurance record in space.[20]

The first job for astronauts Pete Conrad, Paul Weitz, and scientist-astronaut Joe Kerwin was to rescue Skylab following a rough ride to orbit. Part of its solar shielding and one of the rectangular solar electricity panels had been ripped off, and a second solar panel was damaged and wouldn't fully unfold. The first Skylab crew would unlock the panel, and

Astronaut Owen Garriott floats outside NASA's Skylab space station. The orbiting complex was built from parts left over from the remaining Apollo Moon missions cut from the space agency's budget in the early 1970s.

then unfurl a gold solar protective blanket that would flutter like a tarp for the rest of Skylab's life in orbit.[21]

Astronaut Alan Bean would also fly aboard the Skylab station that he helped to pioneer during his days at AAP. He was named commander of the second Skylab flight. Bean would be joined in orbit by crewmates Jack Lousma and Owen Garriott for fifty-six days. Bean believed Apollo-style technology couldn't support a trip to Mars, but Skylab could lay the groundwork for such a mission. "My chief concern as commander," says Bean, "was whether we'd be strong enough after fifty-six days to go outside [on a spacewalk] and get the film canisters from the Apollo Telescope Mount."

Skylab 3 was launched in July 1973. Bean and his crew performed three spacewalks, conducted experiments, made repairs to the workshop, and doubled the length of time that Americans had spent in space on a single mission. The trip also set in motion things like regular exercise to try to minimize the ill effects astronauts feel when returning to normal gravity. "We answered a tremendous amount of questions," says Bean. "Can we go to Mars, which is going to take three-quarters of a year, and be strong enough to walk around? And the answer is yes." It also prompted respect for anyone working aboard a space station or a long-duration spaceflight to Mars. "Going to the Moon was like a vacation," says Bean. "You're working hard, but everything is different along the way. The Moon looks bigger, and the Earth looks littler or vice versa, depending on whether you going out or coming home." Working on Skylab, by contrast, was a grind. He compared it to gearing up for the Super Bowl every day, and trying to do your best work, only to have the same thing repeat day after day. Still, Bean believed his mission aboard Skylab 3 was a greater contribution than his Moon landing during the Apollo 12 mission. "I had more pride at the end of Skylab, and the end of the fifty-nine days in terms of personal achievement," says Bean. "I thought there were more astronauts in the office who could do as well as I did, if not better, on the Moon, than there were who could do as well as I did on Skylab."

Bean's mission on Skylab 3 also pointed out how a rescue mission in Earth orbit would be different compared to saving people on a Mars base millions of miles away. During the initial approach to Skylab, there was a problem with the chemically fueled thrusters used to nudge the command module. The jets would align the capsule so it could dock with the space station. Bean reported seeing what looked like a "sparkler effect" from the thrusters during the rendezvous. Mission control also noticed a pressure drop in those jets. That prompted NASA to question whether it would be safe to use that capsule to bring the crew home. The agency also looked at a rescue flight.[22]

NASA has studied the option of sending up an Apollo capsule to retrieve a stranded Skylab crew. Following the thruster problem on Skylab 3, astronauts Vance Brand and Don Lind began preparing for blastoff in a modified rescue spacecraft. Apollo command modules are built for three people—a commander, a second person to fly the space-craft, and a third crew member. During the Moon landings, that third person was the lunar module pilot. During Skylab, a scientist-astronaut would go. For the rescue flight, two men would pilot the spacecraft in for docking. During the return trip, one of the stranded Skylab crew would sit in the seat between the rescue crew members. The other two space station astronauts would strap into extra seats at the back of the rescue capsule.[23]

Eventually, NASA thought the thruster problem on the command module wasn't serious enough to launch the rescue flight. This kind of mission would have taken days to accomplish. A crew in trouble on the surface of Mars would face a much longer wait for help from Earth. It could take up to six months for a rescue team to fly to Mars. Crew members on the red planet would have to survive on their own until help arrived.[24]

Mars Slips Away

In the face of a shrinking budget, NASA needed to make tough choices on what to do after its Moon landings. If programs like Skylab and the upcoming Space Shuttle were going to survive, something else would have to be scrapped. NASA's post-Apollo Mars mission was the casualty. Walt Cunningham was part of a study group in 1971, prior to the launch of Skylab. Its job was to look at NASA's budget and what was feasible for future missions.[25] "We were looking at alternatives," says Cunningham. "One was a manned Mars mission in 1984." The list also included Sky-lab and the Space Shuttle. NASA didn't have enough money or political will to do everything, so the choice was eventually made to go with the station and the Shuttle. Mars would have to wait.

The disappointments didn't end there.

In 1972, NASA also canceled the nuclear-powered NERVA rocket engine since no Mars trip was being planned. That was a particularly bitter pill for one young scientist who would later influence the future of NASA. One of the people who worked on the engine technology needed to go to Mars was Daniel Goldin. He would go on to lead NASA in the late 1990s and renew the agency's interest in visiting Mars with robotic explorers.[26]

In 1999, as NASA administrator, Goldin's job was to publicize the first Space Shuttle mission led by a woman. Eileen Collins was in training to command Shuttle *Columbia* to deliver the Chandra X-Ray Observatory to orbit. Goldin couldn't help but hearken back to the end of NERVA. "There I was," recalled Goldin, "working on nuclear propulsion to send humans to Mars, only to see the writing on the wall that after the first manned Moon landing, it would be cut." That, in part, prompted Goldin to leave NASA for the private sector.

Even though humans wouldn't venture to Mars after Apollo, the space agency would press on to its most ambitious Mars mission yet. That was Viking and the world's first controlled landing on another world.

ʋiking, nasa's "goʋd bug"

In 1975, the Apollo program ended. There would be no other missions to the Moon for now. NASA's focus had shifted to the upcoming Space Shuttle program. The hope of sending humans on a trip to Mars by building on the rockets, capsules, and industrial muscle that made Apollo possible had faded. Blazing trails to Mars with a manned settlement, and all the hazards similar to those faced by the pioneers of the western frontier, would have to wait.

The robotic Mariner program that paved the way to the red planet was on borrowed time. NASA had set its sights on sending two spacecraft called Voyager on grander trips to the outer planets in the solar system. That task would take far more advanced and reliable vehicles.[1]

In 1965, *Mariner 4* had shattered the dreams of starry-eyed visionaries who hoped Mars would be home to some form of life. Grainy television images revealed the planet to be barren and covered with impact craters like Earth's moon. That mission was followed by later Mariner craft that orbited Mars and took more pictures.[2] However, compared to the ambitious Voyager program, Mariner was looking rickety.

Nonetheless, there would be one last job for the sturdy Mars probes. NASA wanted to land on the surface of the red planet, and Mariner would help the agency do it.

The two robotic explorers could easily have been called "Lewis" and "Clark," but NASA settled on Project Viking. It would generate both excitement and controversy for years afterward because of the dogged effort to search for signs of life on Mars.[3]

Many of the people who labored on the early groundbreaking flights of Mariner had a hand in building and launching the Viking missions. John Casani was project engineer on *Mariner 4*. Back then, it was his job to shepherd the scientists who were building the instruments for the ambitious spacecraft. By the 1970s, Casani had been named project manager for the new Voyager missions. But he still had work to do on Viking. Casani's team would deliver the guidance and power systems for the spacecraft.[4] He knew this would be the final curtain call for the Mariner program, but it didn't leave him feeling nostalgic or weepy.

Engineers don't think like that when a project ends. They just move on to do other things. That's not to say there's no emotional connection between spacecraft and their builders. The hardest part often comes at the beginning of the mission when the vehicle blasts off aboard a rocket at Cape Canaveral in Florida. "It's like postpartum blues after a woman gives birth," explains Casani. During *Mariner 4*, he and his team tinkered with the vehicle on a regular basis for months before liftoff. Experiments had to be positioned on the spacecraft to do their jobs. The gear had to be bolted on and then hooked up to power systems. Then came the Canaveral guys and their rockets, and that was the tough part. "You design it [Mariner], procure the parts, and get it all tuned up," Casani recalled. "Then somebody snatches it from you and shoots it off in a rocket." It would be no different during Viking. As with Mariner, the Jet Propulsion Laboratory (JPL) had to design and build the two spacecraft, not just one.

Viking actually included four vehicles, not counting spare parts. Each spacecraft included an orbiter, which built upon the design of *Mariner 9*. That earlier probe became the first man-made object to successfully enter orbit around Mars. The second part of Viking pushed the boundaries of technology. That section was the lander, which had to enter the atmosphere, protected by a blunt dish-shaped heat shield. It was then supposed to touch down gently using parachutes and retro rockets. NASA had never done this.[5] "The big challenge for the lander was getting through the atmosphere," says Casani. "That [the heat shield] takes out about 75 percent of your descent speed. Then you pop out a parachute to slow down even more. Then during the last bit, your descent rockets take out of the remaining velocity. After that, you put the spacecraft down."

Bruce Murray was concerned as well.

During *Mariner 4*, he led the imaging team that took the pictures of Mars that had generated so much disappointment. Instead of revealing Mars to be a planet of canals and the civilization envisioned by Edgar Rice Burroughs, the television shots arranged by Murray showed the planet to be dead and pocked by meteor craters like the Moon.[6] Since then, Murray had been promoted to director of JPL. Running the spacecraft on Mars was now his responsibility. That included all the technical challenges. "Anybody who wasn't worried didn't understand the problem," says Murray. NASA clearly did comprehend the matter, and this time around JPL would have help. While engineers in Pasadena built the Viking orbiters, space contractor Martin Marietta would build the two landing craft. The whole program would be overseen by the NASA Langley Research Center in Virginia, which had operated in the shadow of JPL during the Mariner missions.[7]

Langley was created in 1917 as the nation's first aeronautical research center. It was home to the world's first pressurized wind tunnel, built in 1922, to test experimental aircraft. That led to groundbreaking work on almost every type of World War II fighter plane that saw combat, including the P-51 Mustang that was made famous by the Tuskegee Airmen.[8]

When the space age dawned, Langley played an early role as well. The facility helped build the stubby, orange-colored Bell X-1 that pilot Chuck Yeager used to break the sound barrier in 1947. Future astronauts like Neil Armstrong and Joe Engle would pilot the black X-15 rocket plane, which would touch the rim of outer space. That craft was also built with the help of engineers and designers at Langley. Before astronaut training became routine at the Johnson Space Center in Houston, Texas, pioneers like John Glenn, Alan Shepard, and Wally Schirra prepared for their Mercury flights at Langley.[9]

In 1968, Mars was the new target, and Angelo "Gus" Guastaferro was assigned as Langley's deputy manager for Viking. "Back then, it was called Titan Mars 73," he recalled. That name referred to the Titan rocket that would carry the vehicle, its destination of the planet Mars, and the target launch year of 1973. Earth and Mars pass close together in their orbits around the Sun like clockwork every other year. Mission planners who want to shoot for Mars generally know when they're going. NASA's first job to get the project going was to secure funding in Congress. The agency believed the first step toward that goal was a name change.

NASA thought "Titan Mars 73" lacked pizzazz. Gus Guastaferro agreed. "The first unmanned missions to the planets were called Pioneer," he recalled. "Because that's what they were. When we moved on to Mariner, if you look at space like an ocean, they [the spacecraft] were mariners. Viking was the conqueror of other lands, although we left out the negative connotation." The name was considered flashy, and it stuck.

However, NASA thought one additional selling point had to be included to win congressional support. Part of the mission had to focus on the search for Martian life. Not everyone agreed that was worthwhile, while others were pragmatic. "I don't think we would have gotten the mission approved if we hadn't looked for life," says Guastaferro.

The argument over Viking's search for living things on Mars has dragged on to this day.

Getting the financial "green light" from Congress didn't mean Viking got a blank check. Money would be tight as Apollo wound down. The Moon program budget was already being trimmed back, and so was NASA's newly renamed Viking program. Mission designers originally wanted each vehicle to be a free-roaming rover that could crawl to more interesting places. Viking was scaled back to two stationary landers, which would be stuck in one spot. The launch was also pushed to 1975. That was OK with many in the Viking program, who believed that even the reduced mission was risky enough.[10]

U.S. spacecraft had whizzed past Mars, and even orbited the planet in the past. But a man-made vehicle had never pierced the Martian atmosphere and landed softly. NASA had put unmanned Surveyor probes on the surface of the Moon. But Mars was different. The red planet had an atmosphere that could incinerate a spacecraft, and stronger gravity, which could mean a crash landing. "It was a wholly new vehicle," says Guastaferro. "You had parachutes and retro rockets. The landers had meteorology labs and cameras with binocular vision. There was a digger to dig soil. It was a laboratory on the surface, unlike Surveyor, which just took pictures." The spacecraft also had to have onboard computers with basic intelligence. If the spacecraft was descending too fast during entry into Mars's atmosphere, each of the two Vikings had to know enough to slow itself down with no help from Earth. Mars was too far away from Earth for mission control to pilot the landers down by remote control.[11] "It was called the six minutes of terror," says Guastaferro.

The finished landers looked like big three-legged bugs. The cameras were mounted inside what resembled small, white grain silos. A black dish antenna was mounted on a mast to send pictures and data to the Viking orbiters, which would relay the signals to receiving stations on Earth. At 1,200 pounds each, the landers were hefty. The whole program wound up costing about $1 billion at that time, making Viking the first of NASA's so-called Rolls Royce missions to the planets. One final complication was in store for builders of the Viking landers. The idea of looking for life on Mars led to the policy that the mission shouldn't contaminate the red planet. NASA agreed to sterilize each of the landing craft to kill off any microbial bugs from Earth that might accidentally hitch a ride to Mars.[12]

NASA space vehicles are typically kept in "clean rooms," where technicians wear coveralls known as "bunny suits" to keep dirt and other contaminants away from the hardware. The plan for Viking was more ambitious than that. Each of the landers was put into an oven and baked at 200 degrees Fahrenheit. The heat was meant to kill off any bacteria or microbes on Viking, but the same process risked frying the electronics on the vehicle. As a precaution, NASA built and delivered a third Viking lander to the launch site at Cape Canaveral. That was in case one of the two primary vehicles was damaged by the cooking. The process had its critics, including Bruce Murray at JPL. "I thought it was pretty nutty," he recalled.[13]

Launch day for *Viking 1* was August 20, 1975, aboard a black-and-silver Titan III rocket. The second spacecraft blasted off the next month. Both would take until the following summer to arrive at Mars. During the long trip from Earth, men like Gus Guastaferro could only sit and wonder about what they couldn't plan for, which was a lot.

The biggest concern was the lack of good reconnaissance of any potential landing sites. The best photos of Mars came from the earlier Mariner missions and the fuzzy images from ground-based telescopes on Earth. The Viking orbiters would scout for suitable landing spots

once the tandem vehicles entered orbit. Still, much of what the landers would face during the trip down was unknown. "You had to pretty
much guess on the environment on Mars," says Guastaferro. "We didn't
know anything about the pressures or winds on Mars. You had to determine how soft the ground was. Nobody had ever been there before,
so we had to guess on how fast to come in."

All of these factors influenced the choice of landing sites for Viking.
Scientists were tantalized by spots like the massive canyon called Valles
Marineris and by the huge extinct volcano known as Olympus Mons.
Both were considered too dangerous. Instead, *Viking 1* would settle on
the gently rolling hills of a region called Chryse Planitia. *Viking 2* would
try for a slightly more hazardous spot known as Utopia Planitia, which
was strewn with rocks that could cause the vehicle to topple over.[14]

Politics Lost

As the *Viking 1* orbiter began circling Mars with the first of the two landers nestled inside, NASA had a political ace up its sleeve for the planned
touchdown. July 4, 1976, marked the U.S. Bicentennial. Mission managers wanted the first unmanned landing on Mars to be the "cherry on
top" of the fireworks displays and the red, white, and blue bunting that
marked the occasion. Attention to the U.S. space program had faded
since the joint mission between America and the Soviet Union the year
earlier. The Apollo-Soyuz Test Project docked the last Apollo command
module with a Russian Soyuz craft in orbit in 1975. This publicity stunt,
born of détente, was long over. Viking was seen as a way to grab some
headlines.[15]

Mars didn't cooperate.

The first landing spot was a region called Cydonia. Scientists favored
it because it was marked with valleys and strewn with rocks and debris.
That area also became part of pop culture because it was home to the
so-called face on Mars, a rock formation resembling a humanlike face,
which prompted speculation of past civilizations on the planet. Still,
Cydonia was judged too dangerous a place to bring down *Viking 1*, so
the July 4 touchdown was canceled. It would take two more weeks before a suitable location, Chryse Planitia, was selected. Its name means
"plains of gold." Scientists who believed Mars had water in its distant

The world received breathtaking photographs of the horizon of Mars from the Viking spacecraft. The robotic probe generated a controversy over experiments to detect the presence of life on the red planet. Courtesy of NASA.

past thought this rolling plain might have been an ancient ocean. They thought it at least might be the result of flowing water from the Valles Marineris canyon. The selection of Chryse Planitia for *Viking 1* didn't make Gus Guastaferro sleep any easier prior to landing.[16]

All seventy of the scientists who helped to design the experiments on *Viking 1* had to take sabbaticals from their universities and head to Pasadena for the landing. That was part of the deal with NASA. The managers and investigators all gathered at JPL in Pasadena and waited. Guastaferro admits he prayed. "What does that say about me?" he asked. "It says I had doubts, and I was reaching out for divine help. The screwdriver I had wasn't long enough to reach the spacecraft if anything went wrong. The only thing I could do was appeal for guidance and support." Providence soon answered, and Viking radioed it was safely on the surface of Mars. Then the phone rang. It was the White House.

Jim Martin was Guastaferro's boss who led the Viking team. He was considered no-nonsense by his staff and anyone who worked with him. "You could come in on Monday and say, 'How about that ball game?'"

recalled Guastaferro. "And he'd say, 'How did that test on Viking go?' That's not to say he didn't have his social side, but how many lights there were in the parking lot or whether we'd get the day after Thanksgiving off didn't interest him." Neither did the phone call from President Gerald Ford. Martin wanted confirmation that the Viking lander was safe and operating before accepting any laurels from Washington. "He left the president on hold," says Guastaferro. "Kind of bold, wasn't it?" Confirmation came, in part, from the first black-and-white photo from the surface of Mars. It was of Viking's own footpad settled into the surrounding soil. The reaction from the scientists and engineers at JPL was a combination of "wow, it worked" and "we were lucky."

Jim Martin then took the call from the White House.

"It [the first photo] confirmed so much," says Gus Guastaferro. "One, we didn't sink into the surface. The imaging system worked. And it confirmed the deep space network could get the data back."

Viking 2 would later touch down at Utopia Planitia. Both vehicles would spend years digging in the soil of the planet and sending back pictures. The last signal from the landers would come in 1982, before they finally fell silent.[17]

Despite the success of Viking, Bruce Murray at JPL had thought all along that mission designers were misguided when they placed such a high emphasis on the search for life on Mars. This same issue would come up again during a time when NASA needed some good publicity very badly.[18]

A Bad Year for NASA Gets Worse

Ten years after the Viking landings, two big stories were making the news in 1986. One was the Chernobyl nuclear plant disaster in Kiev, and the other was the *Challenger* accident. NASA hoped that remembering Viking would give the agency the chance to forget the painful loss of seven astronauts.[19]

The launch of *Challenger* on January 28 that year was one of promise and anticipation. New Hampshire teacher Christa McAuliffe had captured the imagination of schoolchildren across the nation. Each NASA space crew since the 1960s has commemorated its mission by designing an embroidered cloth patch the astronauts wear on their spacesuits.

THE SHUTTLE'S HEROES
A SPECIAL REPORT

Newsweek

February 10, 1986 : $2.00

What
Went
Wrong?

The Future of
Space Flight

Tom Wolfe on
Life at the Edge

Newsweek magazine from February 10, 1986. This issue appeared after the explosion of Space Shuttle *Challenger*, which killed the seven astronauts on board. The crew included NASA's first "teacher in space," Christa McAuliffe. Courtesy of *Newsweek*.

Challenger's patch featured the last names of the astronauts along the border. Next to McAuliffe's name was a little red apple for the teacher.[20] The cargo aboard the Shuttle included a boxy satellite designed to study Halley's Comet, which was coasting within view of Earth when *Challenger* was scheduled to be in orbit.

McAuliffe would be joined inside *Challenger*'s cramped crew cabin by four veteran crew members, including astronaut Ellison Onizuka. His inclusion on *Challenger*'s crew was ironic. His first mission was aboard *Discovery* in 1985. It was marked by a near disaster when one of the Shuttle's solid fuel boosters almost sprang a leak of deadly rocket thrust due to a failure of a gasket called an o-ring.[21]

The boosters are built by stacking segments one on top of another like a tower of soft-drink cans. The point where the barrel-shaped sections come together is sealed with rubber o-rings to keep solid rocket fuel, burning at 6,000 degrees, focused down through the booster's nozzle. Instead, on *Discovery*, the thrust cut into the o-rings. It was the worst example of "burn through" in NASA history, until *Challenger*.[22]

Challenger exploded barely one minute into the 1986 launch, and the crew was lost. NASA was excoriated by an official investigation for management and communication errors that kept the concerns of engineers from getting a fair hearing by NASA leaders. It would be 1988 before the boosters were redesigned and Shuttle flights would resume.[23]

In July 1986, scientists, engineers, members of Congress, and NASA officials gathered at the National Academy of Sciences in Washington, D.C., to forget *Challenger* for a while and celebrate a previous success. It had been ten years since the landing of the first of the two Viking spacecraft on the surface of Mars. The occasion in the nation's capitol allowed for some back- patting and some forward thinking.[24]

Celebrity astronomer Carl Sagan took questions from a group of reporters at the event to remember the Viking landers, which he helped to design and launch. Sagan was perhaps best known as the host of the *Cosmos* television series on public television. The thirteen episodes that first aired in 1980 explained Einstein's theory of relativity and how the stars are born, live, and die. Journalists couldn't help but ask when the show would be broadcast again. His brand of celebrity prompted *Time* magazine to give Sagan the nickname "the showman of science."[25]

Sagan, at that time, had other things to talk about than his enduring notoriety. In 1986, the Soviet Union was still alive, and the United States and Russia were still rivals in space. The central compartment of the *Mir* space station had just been launched months earlier. It would become familiar territory in 1995 when Space Shuttle *Atlantis* docked with the orbiting complex and American astronauts floated inside. In 1986, *Mir* was largely a mystery to people living in the West.[26]

That didn't keep Sagan from speculating about the audacious idea of having the United States and the Soviet Union cooperate on a manned mission to Mars. His proposal would mean constructing a joint space station to build the spacecraft, and robotic reconnaissance missions to

Renowned scientist and astronomer Carl Sagan poses next to a model of NASA's Viking spacecraft. The unmanned vehicle became the first manmade craft to make a soft landing on the surface of Mars. Courtesy of NASA.

the red planet to scout around for landing sites. Another mission, similar to Viking, would touch down on the surface, snatch up a precious sample of Mars soil, pack it aboard a small return rocket, and shoot it back to Earth for scientific study. Sagan was terse when asked why the United States and the Soviets should consider working closely together. Both nations had competed fiercely during the race to the Moon in the 1960s. "If these two countries want to decide that after having filled the planet with sixty thousand nuclear weapons that they wanted to do something on behalf of the human species," said Sagan, "then there would be no more dramatic token of that than a joint manned mission to Mars."

Sagan's distinctive on-air style had been parodied everywhere, from the *Tonight Show with Johnny Carson* to the comic strip *Bloom County*. Still, Sagan was considered more than a pitchman for science. He had made his mark in astronomy years earlier. He theorized that radio signals coming from the planet Venus were due to the broiling 900-degree heat of its surface. He later helped the imaging team on NASA's *Mariner 9* mission, which put the first spacecraft in orbit around Mars. Sagan also helped to pick landing spots for the two Viking spacecraft to visit the red planet.[27] Still, to many TV viewers, he was the science guy in the red turtleneck sweater and corduroy coat who talked about "billions and billions of stars." The opportunity for publicity that a joint Mars mission represented clearly was not lost on him. Sagan thought that every aspect of the trip could be televised to eager viewers on Earth, showing friendly cooperation between space rivals with burgeoning nuclear arsenals.[28] "Mars is a world of wonders," said Sagan, "and a few months of joint or separate exploration, wandering down ancient river valleys or looking for fossils, would be a number of years of sustained scientific excitement and high adventure which could be to a significant extent, doing something of benefit to the human species. I think it could capture the imagination of the world."

Sagan's enthusiasm for the idea of a joint Mars mission was tinged with disappointment about the slow progress at that time to explore the planets. Following the launch of the two Voyager spacecraft, starting in 1977 NASA had halted planetary missions in favor of building the winged Space Shuttles to carry a new generation of astronauts on modest missions in low Earth orbit. NASA, Sagan complained, had basically surrendered exploration of the solar system to the Russians, to Europe, and to the fledgling space program just stirring to life in Japan.[29] Even though he was seriously advocating a manned flight to Mars, Sagan was also among the faction of scientists who wrestled with the drama of sending human beings on missions in space instead of less costly robots to do much of the same work. "To the extent that I'm wearing my science hat," Sagan admitted, "robotic missions are the way to go. They're much cheaper, you can send them into much more hazardous conditions, they don't talk back, they don't have entertainment requirements. And we've seen the spectacular results of the Viking and Voyager missions."

Prior to the success of the *Cosmos* TV series, Carl Sagan had also become the "pied piper" of the science world when it came to the possibility of life on other planets, including Mars. During one episode of his *Cosmos* program, Sagan speculated on what kind of life could evolve on a gaseous planet like Jupiter, where there's no solid surface to walk or crawl upon. He painted evocative pictures of hydrogen-filled life-forms shaped like blimps that gathered lazily in huge herds and floated in the thick atmosphere. Sagan's interest in extraterrestrial life led him to successfully advocate focusing Viking on the search for life on Mars.[30]

Ten years later, in July 1986, that search was about to bite NASA.

Life on Mars? Been There, Done That

Gilbert Levin was no stranger to NASA in the search for life other than that found on Earth. In the early 1960s, he designed an experiment called "Gulliver" to identify microscopic life on other planets. He was also invited to submit plans for an experiment on the yet-to-be-built Viking mission to look for evidence of life on Mars. His plan was to set up a buffet for whatever bugs might inhabit the soil of the red planet. The paper he would deliver during the Viking tenth anniversary celebration at the National Academy of Sciences in 1986 would claim that his device on Viking did find evidence of life on Mars.[31] That was an opinion that NASA treated with respect, but did not share.

Designing a space mission is like packing for a vacation. Everything has to fit in your suitcase or the trunk of the family car. There are size and weight limitations on each spacecraft, and if one experiment makes the trip, another is left behind. Gilbert Levin's proposal was called a "labeled release" experiment to look for signs of Martian life.[32]

The device was based on the notion that living things generate gases as they eat and go about their business. This can be in the form of respiration or the release of gases through digestion. After their arrival on Mars, the two Viking landers, each with a version of Levin's labeled release experiment, went to work. The test called for using Viking's mechanical arm to scoop up some soil and deposit it into a small chamber containing a liquid nutrient laced with radioactive carbon. Levin theorized that if there were any microscopic organisms in the soil, they would eat the nutrient and release gases. The labeled release experiment included a small sensor designed to sniff for radioactive carbon in the thin Martian air above the nutrient dish. Its job was to search for the radioactive burps of microbes.[33]

What happened during the Viking mission became the stuff of intense debate. The sensor detected radioactive gases, which would seem to be clear evidence that something alive was in the Martian soil. "The experiment had an excellent pedigree," says Levin. "It had been approved by four NASA advisory committees that selected experiments to look for life from all over the world." That, Levin insisted, included the criteria for a positive result indicating the presence of life on Mars.

In the years following the mission of the two Viking landers, Levin's chief critic was the late Gerald Soffen, the lead project scientist on Viking. "We did not find life," he stated during an interview in 1986 following the delivery of Levin's paper at the National Academy of Sciences. "I looked at the results, and he looked at the results," said Soffen. "There are just two people with two different opinions. I don't think his case is strong enough."

The difference of opinion was due to a conflicting interpretation of Levin's findings, and because of the results of a second life experiment on Viking called the gas chromatograph mass spectrometer, or GCMS for short. Instead of sponsoring a feast for microscopic Martians and sniffing for their breath, the GCMS scanned the soil of Mars for any trace of organic material that could point to the presence of life.[34]

The results were considered negative.[35]

That led scientists like Gerald Soffen to believe that some kind of chemical reaction in the soil caused the gases that were picked up by the sensor on Levin's experiment. During his 1986 interview, Soffen didn't discount the possibility that a future experiment might find life on Mars, but he thought that the results of Levin's labeled release test weren't convincing enough. "It's like looking up at the sky and seeing it's overcast," said Soffen. "One scientist says he thinks it will rain that day and another says it won't. You won't know until it rains or until the Sun goes down. It's the same in this case." NASA's official opinion was that presence of hydrogen peroxide in the soil likely caused a falsely positive reaction in the labeled release experiment.[36]

That was a view Gilbert Levin declined to accept.

After the 1976 Viking mission, Levin spent ten years trying to duplicate the conditions that NASA concluded had occurred on Mars. That included taking soil samples from the spot on Earth that most closely resembled Mars, the continent of Antarctica. He put those samples into a copy of the labeled release device like the ones on Viking. Gases from microbes were detected. What encouraged Levin even more were follow-up tests where the Antarctic soil was put into the same kind of GCMS that his critics had used to dispute his findings. The Antarctic soil, which contained microscopic life in Levin's test, revealed no organic material when it was scanned by the GCMS. In Levin's mind, if the spectrometer couldn't find organics in Earth soil where life was abundant, the negative results from Mars lacked credibility. "The results of our tests continued to be 100 percent consistent with the likelihood that we had detected living organisms in the soil," says Levin. "The GCMS wasn't sensitive enough to detect organic material in the soil. We could detect organics, and the GCMS could not."

Levin gathered further evidence he believed would back his opinion by enlisting the aid of ground controllers on the *Mariner 9* spacecraft still in orbit around Mars. Critics of the labeled release experiment thought that the presence of hydrogen peroxide would lead to the same chemical reaction Levin believed was the result of living things. *Mariner 9* carried a device called the infrared spectroscopy experiment, or IRIS for short. Levin believed that if hydrogen peroxide was on Mars, IRIS could find it.[37] "No signal for hydrogen peroxide was found," Levin says.

Gilbert Levin delivered his findings at the National Academy of Sciences in 1986. Gerald Soffen, who led NASA's science team during the Viking mission, remained unconvinced. "If there were new facts we'd be happy to consider them, but there are no new facts," Soffen said shortly after Levin's address. "If you're asking if scientists are happy with the results? No, they wish it would have turned out differently. It would have been interesting. But we have to accept nature the way it is, not the way we'd like it to be."

Levin went away from the conference identifying with astronomer Nicolaus Copernicus. The Polish cleric had proclaimed in the mid sixteenth century that Earth was not the center of the universe. That was a view that was not wildly popular at the time. Levin believed that concrete proof of life on a planet other than Earth would be as traumatic on a sociological, philosophical, and even theological level. Gerald Soffen wasn't the only scientist who rejected Levin's conclusions, which Levin found frustrating.[38] "What I object to is the statement from many scientists that we don't have evidence," he says. "The evidence is very strong. If you send an experiment that everyone agrees to and you get a positive result, you can't deny that's evidence." Still, he admits that the issue of life on other planets is one that cuts deep in the scientific community.

Just as Gilbert Levin felt trapped by the adverse reaction to his experiment, Bruce Murray felt trapped that Viking dealt with the issue of life at all.

After leading the imaging team on *Mariner 4* in 1965 that dispelled the myth that Mars was brimming with life, Murray was promoted to director of JPL in April 1976, three months before the first of the Vikings touched down on Mars. He later founded the Planetary Society in 1980 with his friend and colleague Carl Sagan. "You can disagree with friends without having that impact your friendship," says Murray. One of those things was the search for life on Mars.

Sagan came up with the idea of sending greeting cards to alien beings on four spacecraft destined to sail out of the solar system. In 1972, *Pioneer 10* blasted off from Cape Canaveral carrying an engraved gold-plated aluminum plaque. The images included that of a nude man and woman standing in front of an outline of the spacecraft and its dish antenna, giving whoever views the message an idea of what humans look like and how tall we are. Other etchings include a map of where the Sun is located in relation to fourteen neutron stars called pulsars in the Milky Way galaxy. *Pioneer 11* carried a similar plaque in 1973.[39]

Sagan advocated a more complex message on the two Voyager spacecraft set for launch starting in 1977. Each probe carries an etched gold record containing photographs of the people of Earth, music ranging from Beethoven to rock-and-roll legend Chuck Berry, as well as messages from dozens of world leaders. Its protective case contained instructions for whoever found the spacecraft in the distant future to play back the greetings from Earth.[40]

Sagan also pushed for the life experiments to be included on the Viking landers. Bruce Murray thought his friend and his supporters in the scientific community simply jumped the gun. "Have you ever read 'The Gold Bug'?" asked Murray, referring to the Edgar Allan Poe story of a man who goes insane after being bitten by a gold scarab beetle. That prompts him, his best friend, and his manservant to go on an obsessive quest for lost treasure. Murray believed that the single-minded quest by Sagan and proponents to search for life on Mars during Viking was much the same situation. "There was that 'Gold Bug' attitude," says Murray. "The idea of finding life on Mars was so powerful, so seductive, there was a tendency to distort scientific views. It was a pretty heady brew." Striving to find life on Mars carried a greater price than the apparently negative results from the experiments on Vikings. It would be twenty years before another U.S. spacecraft would explore the surface of Mars. Murray believed the agency lost a critical opportunity to learn about the makeup of the red planet during Viking. It had been eleven years since *Mariner 4* coasted past Mars, and now the first of two NASA spacecraft were on the surface of Mars. Viking sent back panoramic views of the rocky surface of two regions of the planet, Chryse Planitia for *Viking 1* and Utopia Planitia for *Viking 2*. Bruce Murray was a geologist by training. With the scales tipped toward the search for life on Mars, Murray felt like a youngster with his nose pressed up against the window of a candy store. He was so close to unlocking tantalizing secrets about Mars, yet so far away.[41]

If astronauts ever visit the surface of Mars, the great distances and isolation will create challenges not seen since the days of pioneering the Old West. Bruce Murray thought that this trail blazing should have begun with an investigation by Viking. He compares it to what the Lewis and Clark expedition did in the nineteenth century. "They were probably more concerned about what the climate was like," says Murray. "They weren't interested in the more sophisticated questions. I think with the earlier Mars missions, there was so much enthusiasm for the search for life that they skipped over the kinds of questions answered by Lewis and Clark." The presence of experiments to look for life had pushed more basic equipment off the Viking landers. First on Murray's wish list was a device called a mass spectrometer, which could have studied the rocks and soil on Mars to determine their chemical composition. That same apparatus could have also studied the atmosphere of Mars. Murray believed that effort was shortchanged by the push to determine if there was life on Mars.

The Viking program seemed all the more miraculous considering how long it took NASA to repeat the act of landing a robotic spacecraft safely on the surface of Mars. The agency tried it in 1999 with a follow-up spacecraft called the *Mars Polar Lander*.

It never made it.

It would take until 2008 for NASA to duplicate the controlled landings of Viking. Prior to the successful arrival of the *Mars Phoenix* spacecraft in 2008, NASA used a more low-tech method of putting vehicles on the red planet.

The agency used beach balls.

the twenty-year gap

Scientists yearning to explore Mars blamed the two Viking landers for a twenty-year lag before NASA would visit the planet again. The space agency had spent an estimated $1 billion on Viking, which would equal $3.5 billion in today's economy. Most scientists agreed that Viking's quest to find evidence of life on Mars had come up empty. It would be 1992 before a plan would be drawn up to resume launches to the surface of the red planet.[1]

The early expeditions to conquer the Old West concentrated on learning the basic "lay of the land." Bruce Murray at the Jet Propulsion Laboratory (JPL) had complained that Viking didn't do this because it searched for life at the expense of studying the geology of Mars. That lesson apparently sank in at NASA. The agency's new administrator, Dan Goldin, ushered in the era of smaller, cheaper, and faster missions.

Instead of looking for Martians, studying the geology of the planet was near the top of the list. The days of the "Rolls Royce" projects like Voyager and Viking were coming to an end, with only a couple of exceptions. One was the *Cassini* spacecraft, which would fly to the ringed planet Saturn and which would have a direct impact on the agency's renewed ambitions to visit Mars.[2]

NASA spacecraft had flown past Saturn previously. The two Voyager probes and *Pioneer 11* had conducted flybys of the planet in the 1970s, snapping photos as they sailed past. The next step for the agency was to position a vehicle there to conduct an ongoing visit. Doing so would pave one scientist's way for a robotic trip to Mars.

An artist's conception of the deployment of the *Galileo* spacecraft from Space Shuttle *Atlantis*. The main antenna on the unmanned probe was stuck partially shut, reducing the amount of data that could be transmitted to Earth. *Galileo* also carried a small probe designed to study the gaseous atmosphere of Jupiter after being dropped from the main vehicle. Courtesy of NASA.

The *Galileo* craft was launched from Space Shuttle *Atlantis* in 1989 on a trip to Jupiter. It settled into orbit in 1995 and spent eight years photographing the brightly color bands of boiling gases encircling the planet.[3]

Saturn was next in line.

The European Space Agency and NASA had started talking about launching an unmanned mission to Saturn in 1984. The initial plan called for Europe to build a small probe to parachute down to the surface of Titan, Saturn's largest moon. The planet-size satellite had intrigued scientists for centuries. It was the only moon in the solar system with an atmosphere. It was also believed to have alien oceans of liquid methane.[4]

The smaller probe would hitch a ride from Earth to Saturn on the main spacecraft, which NASA would build. Originally, mission designers tossed in additional goals of visiting a comet and an asteroid before zeroing in Saturn. The proposed spacecraft was dubbed *CRAF-Cassini*. CRAF was short for "comet rendezvous, asteroid fly-by," and the name Cassini was in honor of the seventeenth-century Italian astronomer Jean-Dominique Cassini, who discovered several of the moons of Saturn and who also used early telescopes to uncover a gap in Saturn's rings that bears his name.[5]

A review of NASA's budget in 1992 showed that the agency's goals exceeded its bank account, so the comet and asteroid visits were canceled. However, scientists would get their chance to go to Saturn and send a tiny probe down to Titan. *Cassini* would be teamed up with the smaller moon explorer, which became known as *Huygens*.[6] It was the namesake of Dutch astronomer Christiaan Huygens, who discovered Titan in 1655.

Signatures from Earth

Some of the signatures of schoolchildren that were digitally recorded on a disc to be included aboard NASA's *Cassini* spacecraft bound for Saturn. The unmanned probe would drop a miniature vehicle called *Huygens*, which would plunge into the atmosphere of Saturn's moon Titan. Courtesy of NASA.

Saturn or Bust

The *Cassini* spacecraft had a boxy main body, topped with a dish antenna to transmit data to Earth and receive commands from mission control. It resembled a twenty-foot-tall birdbath covered in gold foil. The *Huygens* probe nestled against its side, covered in a dome-shaped heat shield, which would protect it during the perilous plunge through Titan's atmosphere. The smaller probe's instruments would study the gases in the moon's smoggy atmosphere and photograph its surface.[7]

Designing and building *Huygens*'s camera would bring Peter Smith into the space business, and set him on a course toward Mars.

Smith worked at the Lunar and Planetary Lab at the University of Arizona at Tucson, after earning his master's degree in optics. He was hooked on space exploration, but he would have to develop extra skills to put a camera on *Cassini*. Smith and his mentor, Martin Tomasko, received NASA funding for the *Huygens* camera in 1989. Then Smith found himself riding herd on Martin Marietta, the company that would build the hardware. "It was the first time I'd ever handled a $30 million contract," recalled Smith. "It was a real education on governmental relations, and kind of overwhelming."

The camera, called a descent imager, was built into the *Huygens* probe. Once *Cassini* settled into orbit around Saturn, the smaller craft would detach for a sixteen-day solo trek toward the cloudy moon. After entering Titan's atmosphere, *Huygens* would descend by parachute for as long as two hours before settling on the surface. Smith's camera would snap pictures as the vehicle dangled from its chute. What it saw during the trip down would depend largely on luck. The camera couldn't turn on its own to snap more interesting pictures. It would take shots of whatever got in the way.[8]

Cassini blasted off aboard a Titan-IV rocket in 1997. The trip from Earth to Saturn took a roundabout path. No launch vehicle was strong enough to boost the six-ton spacecraft directly to Saturn. Instead, *Cassini* brushed past the Sun, Venus, Earth, and Jupiter to utilize their gravity to gain speed for the slingshot trip to Saturn. The voyage was so long that Peter Smith would celebrate a major success, and mourn a major failure involving Mars, before *Cassini* reached its destination.[9]

One by-product of the development of the *Cassini* camera was the spare parts that helped Smith join a team of scientists who would build a spacecraft bound for the red planet. He would team up with another colleague, Dan Britt, who would act as project manager for the camera on NASA's new Mars spacecraft, called *Mars Pathfinder*.

Mars or Bust

NASA gathered scientists and engineers together for a conference in San Juan Capistrano, California, in 1992 to talk about renewed exploration of Mars. It was part of the so-called Discovery missions that were part of the new "smaller, cheaper, faster" philosophy. *Pathfinder* would follow that plan. Viking cost billions; *Pathfinder* would get by on a budget of $150 million. It was chump change by comparison.[10]

The way things typically work at NASA is that a science definition team decides what a future space mission will do. Then the agency takes bids and proposals on the equipment needed to meet those goals. On *Pathfinder*, NASA wanted to send a spacecraft to Mars to study the chemical makeup of dust and rocks on the planet, photograph the surface, and take readings on the weather. The vehicle would include a landing base and a twenty-pound rover, which would roll to interesting rocks for an up-close analysis. The whole package had to be launched on a rocket, reach Mars, and conduct its mission without busting its budget.[11]

Dan Britt didn't grouse about the spartan funding for *Pathfinder*. Before studying to be a planetary scientist, he had worked as an economist for Boeing. Even after Britt made the transition from bean counting to science, he could appreciate the benefits of the "smaller, cheaper, faster" idea. There were the money-saving aspects for NASA, but it also helped scientists desperate to get their instruments into space. Cheaper missions flew more frequently with fewer experiments. That meant more tickets to space on more missions. That, in turn, could add up to more slots for science experiments.[12] "If you have more missions, that cuts the risk," says Britt. "If you lose one mission it's no big deal. If you lose one of those big missions, you're dead for a decade." *Mars Path-*

The *Mars Pathfinder* rover, known as *Sojourner*, crawled off the first NASA lander to touch the surface of the red planet since the Viking missions in 1970s. Courtesy of NASA.

finder would include the camera, a rock spectrometer, and a weather station, and that was it.

The expectations for *Mars Pathfinder*'s camera were as specific as the other parts of the mission. NASA wanted the imager to be able to turn to track the movements of the rover. That way, scientists would know what rocks the rover was approaching and whether an upcoming target was worth studying. Putting a camera on a movable mast was easy on Earth, but Mars presented greater challenges. "The problem is you're in the middle of Mars, and Mars is a cold, dusty place," says Britt. "If your camera swivels back and forth, it's cold there, and most lubricants freeze." That would be one of the many challenges the imaging team would have to face.

Peter Smith and Dan Britt and their colleagues took the spare parts and experience from building the camera on *Cassini* and reworked it for *Mars Pathfinder*. "We went through the proposal in three weeks," recalled Smith. "We started with drawings on a napkin. The concept was very simple. We just put the *Huygens* camera on a mast." Both men knew from past experience that NASA was more likely to approve a plan with proven hardware rather than risk millions of dollars on a design that hadn't been tested in space. Playing the *Cassini* trump card paid off handsomely. "We came out of the blue and blew them away with what we had," says Britt. "We had European partners who were donating most of the parts, so we had double the instrument for what our competitors could do."

Foreign participation on *Pathfinder* wasn't unique to the camera. The Soviet Union had tried and failed on numerous occasions to put a lander on Mars. The Russian space agency developed an instrument called an alpha proton x-ray spectrometer to study the chemical composition of rocks and soil. One flew on the *Vega* mission to the planet Venus, and another was aboard the *Phobos* craft that tried unsuccessfully to reach one of Mars's two moons. A spare spectrometer would be donated to *Pathfinder*.[13]

The camera proposal was submitted in 1992. After a site survey and months of consideration, the plan was approved in 1993.

Now Smith and Britt's team had to make its camera work on a shoe-string budget. The two knew they were working in "economy class" when they saw where they would be working. Building and testing the *Mars Pathfinder* would be done at JPL in Pasadena, California. It occupied one floor of a building built specifically for Viking in the 1970s. "Four billion dollars is a nice thing to have," quipped Britt of the earlier and well-funded Mars mission. *Pathfinder* would get by with less. The way the project was structured was also different compared to Viking. In the past, there would be a design team that would pass its work on to a construction team, which would build the spacecraft. After that, the hardware would be given to a testing group, then the launch team, which would pack it aboard a rocket for launch. Once the spacecraft blasted off, another team would handle descent and landing or orbital insertion. Finally, mission managers would operate the vehicle and conduct the experiments. Observers noticed that problems in one phase of development would be passed along to the next group to solve. That wouldn't happen on *Mars Pathfinder*. "The big innovation with *Pathfinder* was that NASA hired a bunch of guys and said, 'You're responsible for this thing until it shuts down,'" says Britt. "You don't pass your problems along to somebody else, you solve them."

The plan called for the lander to touch down on the Martian surface folded up, then three metallic petals would open like a high-tech flower to expose the instruments, the camera, as well as the rover, which would roll off to explore on its own. Prior to liftoff, the camera required testing to make sure it could successfully photograph rocks on Mars with different-colored filters. The imaging system was two cameras in one for a stereoscopic view packed in an instrument the size of a beer can.[14] Each camera had a filter wheel that rotated to take pictures in twenty-four colors and in infrared. "The color of rocks is diagnostic," explained Britt. "It tells you the composition of rocks. Green ones are made of one mineral, and red rocks are something else."

Even though the *Pathfinder* camera was based on the device aboard the Saturn-bound *Huygens* probe, the team wanted to make sure it would work on Mars as advertised. To do that, the *Pathfinder* camera was set up in a room to take test photos to calibrate the device. A table was positioned within view of the camera and covered in newspaper. A black velvet blanket was spread on top, and various rocks were laid out. The samples were photographed with different-colored filters. The camera was designed so the filters would alternate colors. For example, when the green filter was over one lens, the infrared filter was over the other. That's when the camera team was reminded of a quirk of infrared photography.[15] "You could read the newspaper through the velvet with the infrared filter," says Britt. Simply put, the quirk was that infrared cameras make certain fabrics transparent, especially synthetic ones. That led to an unspoken dress code during the testing process at JPL, especially when the press was invited to cover the story. "I would inquire as to what they were wearing," says Britt with a grin. "If they were dressed out in rayon, I would suggest that they keep that cotton jacket on." This was particularly important if someone planned to stand within view of the cameras. The point was to prevent embarrassing photos that revealed more than expected.

The testing process wasn't the only hurdle the imaging team had to clear. Next came the "gray beards." That was the nickname for elder scientists at JPL who would hold regular meetings to review the progress and ideas behind *Mars Pathfinder*. These veterans were the driving force behind the now legendary Viking project. Now they would sit in merciless judgment of their younger colleagues.

The process meant setting up long tables. The "gray beards" sat on one side, and the *Pathfinder* crew, including Dan Britt, sat on the other. "It was sort of like being in a 'star chamber' in the Middle Ages," says

Britt. "Evidence discovered through torture may be admissible in court, that sort of thing." Viking program manager Jim Martin led the meetings. His no-nonsense management style was legendary inside and outside the Viking team. From the perspective of the *Pathfinder* team, his physical appearance just added to their discomfort. Martin was six foot four inches tall with a flattop haircut that was considered imposing in the extreme.[16]

On the one hand, the *Pathfinder* team wanted to go their own way, and yet the accomplishments of the Viking team were impossible to discount. "There was a huge generation gap between the Viking and *Pathfinder* guys," says Britt. "But there was this awe over what they'd accomplished." Still, the "new kids" knew they were dealing with a different budget, a different mission, and a different time. That still didn't head off significant disagreements on how *Pathfinder* should proceed. The "gray beards" weren't convinced that a camera was needed on the mission, which left Dan Britt concerned. Another bone of contention was the revolutionary plan for landing *Mars Pathfinder* on the surface of the red planet.[17]

The team wanted to use airbags, which would inflate during the final seconds of descent. The cushions would encircle the spacecraft, allowing the vehicle to bounce and roll safely to a stop. It was very different from the orderly combination of parachutes and retro rockets that enabled Viking to make the first soft landing on Mars. That prompted dissension among the "gray beards," but the *Pathfinder* team liked the airbags just fine.[18] "It solved a lot of problems with finding a rock-free landing spot on Mars," says Peter Smith. "If you have thrusters, you can't have any rocks in the way or you'd destroy the lander. With airbags, you'd bounce right over them."

That was the theory, but testing the bags under Martian conditions was difficult on Earth. Mars has gravity only one-third as strong and a much thinner atmosphere. A test chamber at Wright Patterson Air Force Base in Ohio could at least duplicate the atmospheric pressure of Mars. The bags bounced much higher than the *Pathfinder* team expected during the tests, but it didn't shake their confidence. "The airbags turned out to be so strong you could wrap yourself in them and rob a bank," says Britt. "They were three-ply Kevlar, they were really tough." Mission managers on *Pathfinder* thanked the "gray beards" for their input and decided the flight would proceed with the camera and the airbags.

That air of certainty may have worked within the "star chamber" at JPL, but the team would face its true test during the upcoming launch in 1996.

The Blastoff

Peter Smith had designed cameras that would fly on spacecraft to both Mars and Saturn, but he was a rookie observer at the Cape Canaveral Air Force Station. *Mars Pathfinder* was perched on a $50 million Delta-II rocket for liftoff. Smith, Britt, and their colleagues gathered in Florida for the launch. Four years of work would rely on the Delta team, and all the *Pathfinder* crew could do was fly to Brevard County and hope everything went OK. *Mars Pathfinder* was packed inside a protective shell shaped like small, squat version of an Apollo Moon capsule.[19]

Britt had heard about the reliability of the Delta II rocket. The vehicle had a 96 percent success rate, which he didn't find comforting. It meant a 4 percent chance that a spectacular failure could occur. "Would you get on a passenger jet with a 96 percent success rate?" asked Britt. "Hell, no." Peter Smith tried to make the best of it. JPL had rented an apartment near Jetty Park in Titusville, Florida. Veteran viewers of Space Shuttle launches considered that prime territory for rocket watching.

Liftoff day takes on a carnival atmosphere in Brevard County. Spectators drive in from all over and pack hotels and restaurants for the big show. Hardcore rocketry fans arrive early in recreational vehicles to secure prime viewing spots. They hold cookouts and swap stories while waiting for the countdown to reach zero. Vendors hawk T-shirts and ball caps emblazoned with the spacecraft to be launched.

Mars Pathfinder was set for liftoff in December 1996. That meant the temperature would be crisp and mosquitoes would be mostly absent. Unfortunately, the Delta rocket had to lift off early in the morning to put the probe on the right path to Mars. Adding to the tension was that the launch had been delayed two times in a row.[20] Dan Britt waited each night out at Jetty Park, while Peter Smith kept vigil at his rented apartment. "The first night we had a big party," recalled Smith. "Then the second night we had a party, and the third night we had another party." On December 10, 1996, Smith's group gathered on the balcony around 2 a.m. and made bets on whether the Delta would blow up. The only other launches Smith had ever seen were Aerobee rockets, which resulted in a bang and a quick exit. The blastoff of the Delta rocket, with its mightier engines, was a bit different. "It was like the sun coming up. There was no noise at all. You could read a book by the light," recalled Smith. "About thirty seconds off the launchpad, there was the sound, and you felt your chest compress. I had never seen a launch like that." The Moon was visible that night, and, appropriately, so was Mars. "The rocket went out over the Atlantic," Dan Britt remembered. "Then it made a course correction and went straight toward Mars. We all went 'Wow!'"

Mars Pathfinder was on its way.

A successful liftoff didn't make life any easier for Dan Britt and Peter Smith. After a six-month voyage, the probe still had to go through atmospheric entry, the perilous descent, and landing.

Scientists and schoolchildren were all waiting to see if the spacecraft would make it successfully to the surface of Mars. The Mattel toy company tried to cash in on the excitement over the *Pathfinder*'s six-wheeled rover by issuing "Hot Wheels" versions of the lander and the six-wheeled rover. The little robot had been named *Sojourner* after abolitionist Sojourner Truth. The vehicle was set to touch down on Mars on July 4, 1997, but fireworks were the last thing the *Pathfinder* team wanted to see on Independence Day that year.[21]

Peter Smith waited at the mission control center for *Pathfinder* at JPL in Pasadena. JPL had been the control center for Mariner and Viking, and now it was *Pathfinder*'s turn. The team knew exactly what would signal success or failure during the landing. The spacecraft would tell them. The vehicle was programmed to send a "beep" by radio after each major milestone during the trip down. If it survived the fiery trip through the Martian atmosphere, there would be a beep. If the parachutes deployed properly, there would be another beep. If the airbags inflated as designed, beep. If the landing platform opened up and its

An artist's conception of NASA's *Mars Pathfinder* mission. The first lander since the Viking missions in the 1970s included a landing pad with a binocular camera, and a rover named *Sojourner* after abolitionist Sojourner Truth. Courtesy of NASA.

solar panels started gathering electricity, *Pathfinder* would send the most important beep. It would signal that NASA's first spacecraft to Mars in twenty years was in one piece.[22]

Smith and Britt were reasonably certain the landing would work. They had conducted an informal poll of the team members in charge of each phase of the touchdown. They had noticed a trend.

When the engineer in charge of *Pathfinder*'s heat shield was asked how he felt about the landing, he responded that entry into the atmosphere would go fine. However, he wasn't sure about the man next to him who built the parachutes. When the parachute engineer was asked about the chances of success, he said he was positive the chutes would perform as expected, but he was concerned whether the heat shield would protect *Pathfinder*. Likewise, the team in charge of the airbags was confident the shock-absorbing balloons would work fine, assuming the parachutes and heat shield guys did their job. All of the uncertainties made Dan Britt feel pretty good about the upcoming landing.[23] "Everybody we talked to was confident his part of the mission would work," says Britt. "I went home and slept pretty well." And that's just what he did. His first shift at JPL would be hours after the landing, so Britt would wake up and would watch reruns of the touchdown on CNN like everybody else.

Peter Smith didn't have it that easy. He was the man in charge of the camera that would send the first pictures from the surface of Mars in two decades. He was on the hot seat, and he knew it. On landing day, he was standing next to NASA administrator Dan Goldin. All the beeps had occurred to signal each phase of the landing. *Pathfinder* was down, but there would an uncomfortable time lag before it would radio the first picture from the camera Smith helped to build. "I remember all these high-level NASA people staring at me," says Smith. "Their eyes seemed to say, 'This had better work or we're going to roast you.'"

The sweating stopped when a huge Mars rock appeared on the television screens at JPL. It was six feet long and four feet tall, and the assembled personnel thought it looked like a bear. They nicknamed it "Yogi." The crowd in the control center went wild. "It was like they'd never seen a Mars rock before," says Smith. A later panoramic shot showed the opened petals of the *Pathfinder* lander, the *Sojourner* rover, and twin Martian peaks off in the distance.[24] Later, the biggest shock for Dan Britt was that the red planet wasn't really red. "I thought it was going to be red," says Britt. "It turned out to be a yellowish orange, which is OK." The camera's view was also higher off the ground than the imaging team had expected. *Pathfinder* had settled on top of its own deflated airbags, which propped the vehicle a few inches higher off the ground than mission designers had anticipated.[25]

The Little Rover That Could

The *Pathfinder* team didn't have much time for celebration. The vehicle on Mars was simply a demonstration project for future missions. That meant there were few redundancies built into the spacecraft and the rover to keep them operating under the grueling conditions of Mars. *Sojourner* was designed to last only seven days, so mission managers wanted to make every moment count. The six-wheeled explorer began to slowly crawl to a series of rocks with nicknames taken from cartoon characters. Photos taken from the landing-base camera would show the rover dwarfed by the boulder called "Yogi." JPL staff would come up with other names for other rocks like "Barnacle Bill" from the Popeye comic strip, "Flat Top" from Dick Tracy, and "Stimpy" from the *Ren and Stimpy Show*.[26]

Mission scientists were interested in doing more than looking at rocks. *Sojourner's* six metal wheels could be spun in different directions to cut small trenches in the soil. The rover's spectrometer could then be aimed into the grooves to study the different layers that had been exposed. However, the robot couldn't turn "on a dime," and at certain points in its mission it appeared to be at risk of being stuck in clusters of rocks. There was also the concern that the rover's solar panels could be contaminated by the dust kicked up by its wheels. The danger didn't keep people like Dan Britt from suggesting unnerving ideas to the men and women who actually did the driving. "We kept on asking for things they hated to do," Britt recalled. "It was fun to watch them cringe when we came up with these requests." Another challenge for the mission team was shifting their patterns of working and sleeping. Days on Mars are forty minutes different than on Earth, so the daylight hours when *Pathfinder* could be operated kept shifting. Mission managers could be working days at one point and nights at another.[27] Scientists who worked on the lander also had to get used to being upstaged by the *Sojourner* rover, which attracted the most interest from the public and the press. "There was a kind of beauty about that little rover," admitted Peter Smith, "with its little solar panels and its six little wheels. When it struggled to get over rocks, it was like *The Jetsons* and their little robot dog. Everybody could identify with it."

Mars Pathfinder operated beyond its design limits, but contact was eventually lost after three months. The camera on the landing pad delivered 16,000 images, and the rover added more than 500 on its own. A large poster-size print of that early shot of the Martian horizon with the twin peaks in the background is tacked to the wall of Dan Britt's office at the University of Central Florida in Orlando.[28]

The loss of *Pathfinder* was sad for the men and women who worked on it. But Mars and Earth were well on the way toward lining up again in their orbits around the Sun in 1998. There would be yet another opportunity to send a mission. This time, NASA wanted to have two launches. One would be an orbiting spacecraft to photograph Mars, and the second would carry a separate vehicle to land at the planet's South Pole. The timing was so tight that Peter Smith's imaging team was already designing the camera for the *Mars Polar Lander* as plans came together for *Mars Pathfinder*.[29]

The Doomed Mission

The $160 million *Mars Polar Lander*, known by its initials MPL, was intended to resume NASA's search for evidence of life on Mars. It would land by parachute and engine power at the planet's South Pole to dig into the soil to look for signs of ice. It would be the first controlled landing on Mars since Viking in the 1970s.

During the trip down, the MPL was also designed to launch two bullet-shaped microprobes, which would dive toward Mars and hit the planet at 400 miles per hour. The probes were nicknamed *Amundsen* and *Scott*, in honor of two South Pole explorers on Earth. Each of the projectiles was equipped with a heat shield that would shatter upon impact with the surface. Part of the microprobes would drill into the soil, while the upper halves with small antennae would remain above ground. *Amundsen* and *Scott* were intended to radio back data from their experiment packages, including accelerometers and a test chamber, which would do chemical studies on the Martian subsoil.[30]

Like *Mars Pathfinder*, the main landing craft would be equipped with a camera on a mast to photograph the MPL's surroundings. Peter Smith's team at the University of Arizona would design it. The process began two years before *Pathfinder* touched down in 1997. By the time the lander and the *Sojourner* rover had made it to the surface, Smith was putting the finishing touches on the camera for the MPL. "It really stressed the team to deliver a camera and operate another on the surface of Mars," he recalled. "But with the two cycles of landings [on Mars], this was the busiest time I can remember." Smith was even being pressed to come up with imaging equipment for NASA's mission after the polar lander, which was the *Mars Surveyor 2001* flight. As its name implied, the spacecraft would feature an updated version of the *Sojourner* rover, was intended to land on Mars in 2001, two years after the MPL.

The High-Tech Hockey Puck

Dan Britt would make a unique contribution of his own to the MPL mission. NASA wanted to make sure that the photos the polar lander transmitted to Earth represented the same colors humans would see if they were standing on the planet. Britt's solution was an eye chart the size and shape of a hockey puck. It would be covered with one-inch dots of colored rubber to help calibrate the camera on the MPL. When photographers send their film or jpegs to a lab for processing, the technicians can reference what colors generally look like on Earth. Cameras on Mars don't have that advantage because the atmosphere is made of carbon dioxide, and the sunlight is weaker since the planet is farther away.[31] "In order to get the colors right, you have to be looking at something you can reference," says Britt. "You have to be sure what is red, or blue, or green, or gray. That's what I make." Since the mission called for each photo session to begin and end with a shot of the eye chart to ensure color accuracy, Britt's invention was likely to be the most photographed object during the mission of the MPL.

The importance of accurate colors was especially evident following an embarrassing episode during Project Viking back in the mid-1970s. Once the first lander touched down on Chryse Planitia, it took a panoramic shot of the Martian horizon. The series of photos were radioed to JPL to be processed and released to the public as well as top mission managers like Angelo "Gus" Guastaferro. He helped guide the mission from its inception in 1968 to its launch in 1975, and he wanted to see the pictures, too. The cameras on Viking were working fine, so mission control wanted something dramatic to show for it. "It was like the kind of shot you'd take on vacation," recalled Guastaferro. "And we released about a dozen of them." It was after the pictures were issued that something didn't seem right. Someone in the JPL photo lab had used his own judgment in adjusting the colors. "They thought the sky was supposed to be blue," says Guastaferro. Corrected pictures were printed to give a more accurate representation of what Mars looked like. "It turned out we had more of a red sky," says Guastaferro. "Well, not red, more like pink." Some of the inaccurate Mars pictures did get out. Like misprinted postage stamps, the "blue sky" photos are considered valuable by collectors of space memorabilia.

Dan Britt and the imaging team on the MPL would settle for accurate pictures, not collectors' items. The MPL launched on January 3, 1999, aboard a Delta II rocket like the one that propelled *Mars Pathfinder* to the red planet. The second craft, called the *Mars Climate Orbiter*, blasted off the month earlier.[32]

Nine months later it vanished while on final approach.

The MPL was closing in on the red planet while an investigation discovered the loss of the orbiter was due to human error. A NASA report, released in November 1999, stated that contractor Lockheed Martin programmed the thrusters on the spacecraft using English measurements like feet and inches. JPL had been using the metric system for nine years. Instead of calculating the final approach using units of measure called Newton-seconds, the spacecraft was programmed with pound-seconds. The investigation also showed the system of checks and balances at JPL had failed to catch the mistake. The result was that the *Mars Climate Orbiter* came in too low to the Martian atmosphere and likely incinerated.[33]

NASA wanted to clear the air regarding the lost spacecraft before the MPL made its final approach. The spacecraft's programming was double-checked to make sure the same English/metric mistake wasn't made again. The measurements appeared sound.

But something else was about to go wrong.

During the Viking program, the moments before touchdown were described as "six minutes of terror." The final approach for the MPL would live up to that phrase when that spacecraft disappeared as well. The MPL had screamed through the Martian atmosphere as expected. The parachute should have popped free and the engines ignited to cushion the final moments before arrival. All the signs indicated the vehicle had successfully reached its landing spot on the South Pole.

But then the MPL fell silent, just as the *Mars Climate Orbiter* had done back in September.

There was no signal from the spacecraft, and that wasn't the way things were supposed to occur.[34] Peter Smith tried not to worry at first, even considering the painful loss of the previous Mars mission. "People weren't too upset on the first day," says Smith. "There were a lot of ways for the vehicle to go into safe mode." Safe mode is a backup mode if the lander loses touch with Earth. The vehicle protects its systems while transmitting a series of radio signals to locate mission control. That went on for a day, and then the mood at JPL became darker. "After the second day when you got no signals," recalled Smith, "you started to feel defeated, and you knew we'd lost the vehicle. It was misery."

The Aftermath

Losing missions to Mars was nothing new for either the United States or Russia. The apparent failure of the MPL would become more serious when the cause was linked to human error, just as on *Mars Climate Orbiter*. The truth would be laid out during testimony before the Science Committee of the U.S. House of Representatives.

JPL convened an investigation into the double failure of the two Mars vehicles. For his part, NASA administrator Dan Goldin asked retired aerospace executive Thomas Young to lead a similar probe. Young had worked as program manager for Viking in the 1970s and later as director of the NASA Goddard Space Flight Center. In the private sector, he had finished his career at Martin Marietta as an executive vice president after serving as president and CEO. The result of the inquiry, delivered to Congress, wasn't an indictment of Goldin's "smaller, cheaper, faster" philosophy.[35] However, it hinted strongly that *Mars Climate Orbiter* and the MPL had been "nickeled and dimed" to death by a budget that was at least 30 percent too small for the task.

While the orbiting spacecraft had been lost due to a mix-up of English and metric measurements, it appeared that the MPL had been killed by a single line of missing computer software code. Young's team studied evidence leading up to the loss of the lander, and NASA spacecraft in orbit around Mars struggled to photograph the wreckage at the planet's South Pole. The investigation concluded that after the MPL entered the atmosphere, it fired its retro rockets as designed. However, when its landing legs snapped open for touchdown, the jolt apparently sent a false sensor signal to the spacecraft. The vehicle thought it was safely on Mars and shut its engines down. The lander was believed to be 100 feet off the ground when the thrusters were cut off. Investigators concluded the $160 million spacecraft plummeted at fifty miles per hour and crashed.[36]

On April 12, 2000, Young told the House Science Committee that the software on the polar lander should have been robust enough not to be fooled by an errant sensor signal, but it wasn't. "Specifically, software testing was inadequate," Young testified. "Equally important, the navigation team did not understand the spacecraft, and was inadequately trained." The final report also pointed to engineers who had to work eighty hours per week to keep up with the demands of the dual mission and its sparse budget. The reforms sought by the Young commission included more realistic funding, better staffing with experienced managers and sharply defined responsibility, and a pattern called "test as you fly and fly as you test."[37]

Back to the Drawing Board

The first casualty was NASA's 2001 mission called *Mars Surveyor*. It was a copy of the MPL with a robot rover like *Pathfinder's*. Mission designers would have to wait close to eight years before trying again to make a soft landing on Mars like Viking. NASA's next touchdown on the red planet would fall back on the airbags that proved so successful on the *Pathfinder*.[38]

NASA's *Mars Phoenix* spacecraft built upon the failure of the *Mars Polar Lander,* which is believed to have crashed on the surface of the red planet in 1999. *Phoenix* scraped the surface of Mars's polar cap to search for signs of ice and organic material. Courtesy of NASA.

Peter Smith worked on the science team for the Mars Exploration Rovers, *Spirit* and *Opportunity.* The twin robots landed on opposite sides of the planet in 2004. Smith also built the camera for a failed European mission to Mars called *Beagle-2.* He also took solace from his first venture into space technology, the *Cassini* mission to Saturn.

NASA's *Cassini* spacecraft finally entered orbit around Saturn in 2004. The smaller European-built *Huygens* probe, containing Smith's first space camera, was launched toward the atmosphere of the giant moon Titan. The imager, which was the ancestor of the cameras on *Pathfinder* and the MPL, was switched on and put through its paces. One shot of Titan's alien surface sticks out in Peter Smith's mind. "It looked like the coast of Italy," he recalled. "It had these little rivers, and what looks like a dry lake bed with little hills and valleys. It was so Earth-like."

But all this time Smith wanted to put to rest the ghosts of the MPL. He was the one who handed out pink slips to thirty staff members when NASA cut off funding after the crash. Smith and a number of investigators wanted another chance to do the job left undone.

That led to a big box delivered by Lockheed Martin with the leftovers of the *Mars Surveyor* spacecraft. A new vehicle had been created appropriately called *Mars Phoenix*, after the mythical bird reborn from the ashes of its own death. Smith would lead the mission as principal investigator.

The *Mars Phoenix* spacecraft was essentially a reworked version of the canceled *Surveyor* lander. That gave Smith and his team a head start, since the vehicle was mostly assembled. The new *Mars Phoenix* flight had a budget of over $4 billion, which allowed for testing that was absent from the failed MPL. It was an eye-opener for the *Mars Phoenix* team, since a number of fatal design flaws were spotted on their spacecraft, any one of which could have led to another embarrassing failure. The landing radar, which was supposed to give the vehicle its view of the ground during final approach, didn't work. Another flaw might have doomed the spacecraft even earlier.[39] "The separation connectors that separated the cruise section [of the spacecraft] from the lander didn't release at cold temperatures," says Smith. "They had been tested at room temperature only. We had to add heaters to make sure they'd work near Mars."

The Second Go-round

With the testing done, the focus turned to the science on the *Mars Phoenix* mission and the ambitious plan to make the first controlled soft landing on Mars since Viking. A robotic arm would dig into the soil of the planet's North Pole to look for permafrost that could have been hospitable to life sometime in the ancient past. Soil samples would be cooked in small ovens to study their chemical makeup.[40]

Mars Phoenix blasted off aboard a Delta II rocket in August 2007. The launch vehicle was familiar to JPL. Similar rockets had propelled *Mars Pathfinder* on its way, as well as the doomed *Mars Climate Orbiter* and the MPL. The typical cruise to the red planet and final approach were especially tough since the catastrophic loss of the 1999 landing mission was in the back of everyone's mind. "It was very nerve-racking," says Smith. "That included the people at JPL who thought there was only a fifty-fifty chance it would work." The public affairs team that would handle the press was also preparing for the worst. A stack of printed news releases had been written. Each one explained why a different phase of the landing of *Mars Phoenix* had failed. Fortunately, they would not be needed. "It was fabulous," recalled Smith. "The vehicle touched down after using only half the fuel it had on board. The landing couldn't have gone better." A public affairs officer stationed near the stack of doom-and-gloom press releases slowly tore up each one as each phase of the landing progressed successfully. The ripped-up paper was used to shower the mission management team with impromptu confetti.

The biggest impact of the *Mars Phoenix* touchdown for Peter Smith was renewed respect for the people who made the only other soft landing on the planet. They were the Viking engineers who, as the "gray beards," had mercilessly harangued his *Mars Pathfinder* team a decade earlier. "I have the greatest admiration for those guys," says Smith. "Especially considering what they went through in the 1970s. It was an amazing challenge."

In the years following the landing of Viking, astronomer Carl Sagan observed that unmanned missions didn't talk back like astronauts and didn't require food and entertainment on their missions. Robotic flights to Mars that ended in catastrophe also didn't need memorial services. Landing people on the surface of the red planet would be a far greater challenge than *Mars Phoenix*, and NASA had a long way to go before being able to clear this hurdle. Step one would be to move on to to a new spacecraft and to retire NASA's aging fleet of Space Shuttles. It would be a political and painful process.

the shuttle's
long good-bye

NASA's Space Shuttle was considered the most sophisticated flying machine ever conceived. Still, the vehicle's design with liquid-fueled main engines, solid rocket boosters, and an external tank trapped the vehicle in low Earth orbit. The craft couldn't fly higher than the Hubble Space Telescope, much less make it to the Moon or Mars. If astronauts were to ever set out to settle Mars as their ancestors did with the Old West, the Shuttle would have to go. And that process would prove to be a painful one for the people living and working near the Kennedy Space Center.

On June 23, 2008, U.S. senator Bill Nelson of Florida convened a hearing of the Senate Subcommittee on Space, Aeronautics, and Related Sciences in the hearing room of the Port Canaveral Authority near the Kennedy Space Center. The subject wasn't sending astronauts back to the Moon or the hazards of pioneering missions to Mars. The topic that day was more sensitive and more political. NASA administrator Michael Griffin and his staff had been summoned to testify on job losses in Florida following the retirement of the Shuttle.[1]

An earlier estimate, given before Congress in February, put that figure at a jarring 6,000 positions that would be cut. The news sent economic shockwaves through Brevard County, Florida, and political ones through Senator Nelson's subcommittee. "So, the only NASA centers that are taking it in the neck are Kennedy Space Center and Michoud?" Nelson asked during the proceedings. "That is where our contractor losses will be most severe, yes sir," responded Griffin. Kennedy launched the Shuttle, and the NASA Michoud factory, east of New Orleans, built the external fuel tanks used during blastoff.[2]

Following the *Columbia* disaster in 2003, which killed seven astronauts, President George W. Bush set in motion a plan advocated by Michael Griffin. The Space Shuttle fleet would be retired in favor of a new space capsule called Orion. The spacecraft, reminiscent of the gumdrop-shaped command module from the Apollo Moon landings, would be launched on a solid fuel rocket called Ares-1, with a larger, unmanned cargo booster known as Ares-5. The change would put America on a course toward returning astronauts to the Moon as a prelude to far riskier missions to Mars. The plan would mean no need for the hundreds of workers who built the Space Shuttle's big external fuel tanks at the Michoud plant. Thousands more who processed the spacecraft at the Kennedy Space Center for liftoff would lose their jobs as well.[3]

For Brevard County, the situation resurrected ghosts from the end of the Apollo program in the 1970s. Thousands of pink slips went out once Neil Armstrong's footsteps on the Moon were assured. Nelson was a rookie member of the U.S. House in 1971 during the height of Apollo and during the economic plummet in Brevard following the end of the lunar program in 1975. He later flew aboard Space Shuttle *Columbia* as a congressional observer in 1986. The Shuttle landed safely just days before the *Challenger* disaster. For Nelson, space was personal.[4]

Along with the concern over Shuttle jobs, the race for the White House between Republican John McCain and Democrat Barack Obama was nearing its height. Both candidates had made statements in favor of continuing the U.S. space program, but NASA supporters in Congress and Brevard County wanted specifics.

All of these agendas converged at Port Canaveral with about a thousand NASA workers carrying picket signs bearing slogans like "America, the place for space," and "Space equals national security." Instead of the white shirts, neckties, and pocket protectors normally expected for rocket scientists, the demonstrators wore mostly blue jeans and T-shirts emblazoned with union logos. The NASA workforce that maintained the Shuttles was mostly blue collar. It was composed of people ranging from journeyman electricians to graduates of Brevard Community College who studied the process of sticking on the ceramic tiles that protected the spacecraft during each fiery reentry.[5]

Rally organizer Dale Ketchum, who worked at the University of

Central Florida in Orlando, made the phone calls to get the group together. He sat in on the initial meetings with Nelson's subcommittee staff, who didn't think a big venue was needed for the hearing. Meetings about space, they thought, just didn't attract crowds.

Ketchum's response was to create one.

The demonstrators at Port Canaveral posed for a huge group photograph to be delivered to both the McCain and Obama campaigns. "We want to know what their vision for space is, and what role Florida plays in that," said Ketchum. "We're all voters, and we have the right to ask."

The protesters at Port Canaveral were worried about their jobs. Members of Congress were worried about having astronauts in the passenger seats of spaceships built by the Russians. The current retirement plan for the Shuttle would leave at least a five-year gap while NASA built and tested the proposed Orion capsules. During that gap, American astronauts would have to rely on the Russian space program and its Soyuz spacecraft to go to and from the International Space Station. The idea galled both lawmakers and NASA leaders.[6] "It is unseemly in the extreme," NASA administrator Michael Griffin told Senator Nelson's subcommittee in February 2008. "But there is no viable option. We are, today, dependent on the Russian Soyuz for the sustenance of the International Space Station."

Bill Nelson and Republican senator Mel Martinez, also of Florida, were on hand for the subcommittee hearing, and each saw the idea of NASA relying on Russia as both a national security issue and a political one. The demonstration outside the Port Canaveral Authority gave them an opportunity to fire up the crowd. Both lawmakers took the mike at a small stage set up outside within view of the protesters and their colorful signs. "We're going to be laying off people at the Kennedy Space Center," Nelson shouted to his audience, "so we can go hire Russians to build the Soyuz spacecraft, so that the United States government can buy rides on the Soyuz." The crowd drowned out the rest of Nelson's statement with boos. The new Orion capsule could pave the way to the Moon and Mars, but having to go hat in hand to the Russians for rides on Soyuz capsules clearly rubbed U.S. space proponents and politicians the wrong way.

Following the hearing inside, Nelson and Martinez took questions from a gaggle of reporters who zeroed in on whether they intended to lobby the McCain and Obama campaigns to take more comprehensive stands regarding space. "It's terribly important," said Martinez. "Not only to jobs here in Central Florida, but to the nation. And I know Senator McCain's commitment to NASA is there. As a former pilot, his support of manned flight is there." Not to be outdone, Nelson not only pledged to pressure Obama but also to make sure any statement the Illinois senator made about space would be delivered in the shadow of the Kennedy Space Center.

Within weeks, the candidates responded.

On Friday, August 1, one day before addressing the National Urban League convention in Orlando, Obama drove to Brevard Community College in Titusville. He was accompanied by Bill Nelson. Over a thousand supporters crowded the small campus for the rally. "Under my watch," Obama told his well-wishers, "NASA will inspire the world once again, and make America strong, and grow the economy, and make the economy stronger, not just here in Brevard County."

This was a significant change for Obama, and his critics immediately pounced. Previously, the senator had advocated delaying the new spacecraft for five years to provide funding for education. Now, he didn't.[7]

McCain supporters claimed the Democratic candidate was just trying to drum up votes. The Obama campaign said it was part of the senator's plan to reinvigorate the economy. Ironically, observers noted that Obama's plan for NASA closely mirrored President Bush's strategy of retiring the Shuttle and relying on Russian Soyuz vehicles to transport astronauts until the Orion capsule was ready to go. Obama spokespeople claimed it wasn't an endorsement of Bush policy, but rather an effort to make the best of a flawed White House plan. The Democratic proposal also included $2 billion to speed up production of Orion and to conduct one additional Space Shuttle mission. That would delay the end of the program until 2011.[8]

Seventeen days later, John McCain visited the same community college as Obama. Instead of a rally decked out in red, white, and blue bunting, McCain met quietly behind closed doors with about two dozen leaders of Brevard's aerospace community. Participants described it as a strategy session by the Arizona senator to map out how to protect space coast jobs while speeding up production of Orion with $2 billion in extra money for NASA. After the meeting, McCain gave a statement to the press. "I think we must take additional steps to expand the benefits of NASA's research," he said. "We need to harness the innovative technology that's developed on the space coast to find commercial applications that will benefit the economy." At the conclusion of his statement, NBC reporter Kelly O'Donnell asked if he would be taking questions. "No," McCain responded, already halfway out of the room.[9]

Observers noted that McCain's reluctance to rely on the Russians would likely spur a plan to keep the Space Shuttle flying at least until 2011, if not longer. Ultimately, what was expected of the new president were more Shuttle flights to maintain America's ability to send astronauts into orbit, while supplementing NASA's launch efforts with flights on the Russian Soyuz. It was also hoped that there would be money to make Orion available sooner than the 2015 date specified by the Bush plan.[10]

The anguished pleas to save Space Shuttle jobs would also reach the ears of former Martin Marietta president Norman Augustine. He and a panel of aerospace experts would later play a pivotal role in carving NASA's future path once a Democratic victory was sealed.

Barack Obama's triumph at the polls prompted an uncomfortable process at NASA, with the president-elect's transition team reviewing all of the space agency's programs, in a procedure the new administration called "looking under the hood." Both sides denied published reports of friction. During a teleconference to update the press on the Orion program, NASA's deputy associate administrator for exploration, Doug Cooke, told reporters that relations between the agency and the new White House were just fine. "They're asking the same questions we were getting from everyone, which is what they should be doing," said Cooke. "And it's our job to get the best answers we can." Early on, possible tweaks to NASA's path included extending the Space Shuttle program or launching the new Orion space capsule on European-built Ariane 5 rockets. More drastic proposals included merging NASA with the Pentagon's space program, possibly utilizing military Titan-IV or Atlas-V rockets to launch Orion.[11]

Out with the Old, in with . . . What?

As prickly as talks between NASA and Obama's transition team might have been, the agency's ultimate fate began to unfold during meetings of what became known as the Augustine Commission. The presidential panel was the namesake of its chairman, Norman Augustine. He was familiar with the process of crafting lofty goals for the nation's space program. Augustine also witnessed how those plans would be tweaked or brushed aside by the winds of politics.

In 1990, following the celebration of the twentieth anniversary of the Apollo 11 Moon mission, which put the first men on the lunar surface, Augustine was asked to chair a review of NASA's future space goals. President George H. W. Bush named the Advisory Committee on the Future of Human Spaceflight to examine every aspect of what NASA was doing and how that course should be changed in keeping with the times. One major accomplishment for the group was to recommend altering the purpose of the proposed space station *Freedom* to make it a science laboratory and a base for human space operations. Bush's successor, William Clinton, retooled the *Freedom* space station into the International Space Station. Not every idea from Augustine's advisory panel was embraced, however. The group also recommended that NASA's controversial Space Shuttle be replaced with a heavy lift launch vehicle that could eventually pave the way back to the Moon or on to Mars. That idea was ignored, only to be put back on the agenda for President Obama's space commission to consider, again with Norman Augustine at the helm.[12]

The panel, named in June 2009, was officially known as the Review of U.S. Manned Space Flight Plans Committee. When the 1990 space commission gathered, NASA's future was a clean slate of possibilities that could be considered at a somewhat leisurely pace. The 2009 Augustine Commission's job, in contrast, was like reading a road map to plot a vacation route while barreling down the interstate with the kids acting up in the backseat. NASA's Space Shuttle program was destined to end within a year, with the replacement space capsule, called Orion, scheduled to fly in 2015 or 2016. The International Space Station was set to be abandoned around that same time. The plan called for the orbiting complex to be guided down into the atmosphere to safely burn up.

Critics compared the notion to torching the billions of tax dollars spent to build the station and launch the parts, piecemeal, on the Shuttle. The Augustine Commission had to recommend changes while a gun was pointed at NASA's head. Suggestions had to come quickly.

All of the panel's members confessed early on to having "drank the Kool-Aid" regarding the desirability of a vibrant space program. But the board also reflected the changing political and technological world in which NASA now lived. Its members included former astronaut Sally Ride. She helped investigate NASA's two Space Shuttle disasters, *Challenger* in 1986 and *Columbia* in 2003. Her last mission in orbit was in 1984. Leroy Chiao also wore the gold lapel pin afforded to veteran NASA astronauts. However, he represented the current generation of space fliers, having completed a long-duration mission aboard the International Space Station in 2005. He arrived to the orbiting outpost aboard a cramped Russian Soyuz space capsule, which blasted off from a launchpad in the former Soviet nation of Kazakhstan. He returned to Earth in a similar Russian-made vehicle, which was unheard of during Ride's tenure at NASA.

The evolving commercial space industry was also represented on the board. Jeffrey Greason was CEO of the XCOR Company, which wanted to challenge Sir Richard Branson and Virgin Galactic in the field of space tourism. Virgin proposed taking groups of well-heeled customers to the rim of space for a taste of weightlessness. XCOR's ambition was to do likewise, but in a two-seat rocket plane initially piloted by former NASA astronaut Richard Searfoss. Greason raised eyebrows by initially challenging the need for NASA.

The world of politics was also included on the Augustine Commission with the naming of retired air force general Lester Lyles. He was reportedly on the list of possible candidates to be President Obama's nominee for NASA's new top administrator. Subsequent pressure from members of Congress led to the eventual selection of retired Marine Corps general, and former astronaut, Charles Bolden.[13]

The Augustine Commission considered new rockets and destinations far from Earth, but also the human side of the proposed changes at NASA. The board took its show "on the road" and convened at the Cocoa Beach Hilton near the launchpads of the Kennedy Space Center on July 30, 2009.

The commission members sat at tables assembled in a ballroom. Folding metal chairs were set up for the 100 or so members of the public and the media who gathered to hear the reports and testimony. Electrical cables snaked around the walls of the room to accommodate cameras from the NASA TV network as well as reporters who needed to record the proceedings for their stories.

Florida's lieutenant governor, Jeff Kottkamp, took the microphone to lay it on the line. The loss of the Space Shuttle program would hit the Sunshine State harder than anywhere else in the nation. An estimated 7,000 Floridians would be left without jobs during the so-called gap. That was the span of time between the last Shuttle mission and the first proposed launch of Orion. During the gap, there would be no U.S.-owned transportation system to send astronauts to space. "The extended gap from the perspective of the state of Florida is unacceptable, we think, not just for our state but our nation," said Kottkamp. "And if there's any option that can bridge that gap, the state of Florida implores you to give it the strongest consideration." A familiar face from Senator Bill Nelson's hearing at Cape Canaveral in 2008 also spoke when the public was invited to testify. While many of the people taking the mike wore NASA T-shirts and ball caps, Dale Ketchum of the University of Central Florida came in a business suit. "There is no recommendation that you can make that won't result in thousands of jobs lost here," he said. "We recognize that." Both Ketchum's statement and the one from Lieutenant Governor Kottkamp were met with stony silence from the commission. Norman Augustine's mandate from the White House didn't include coming up with any options that supported protecting any part of NASA's workforce. In that instance, he didn't plan to break the rules.[14]

That's not to say the fate of the people who turned wrenches at the Kennedy Space Center didn't figure into the deliberations at all. Augustine said his group would formulate recommendations based on what he termed "areas of merit." "One of those areas of merit is maintaining critical skills," the chairman said. "How much weight the president cares to apply to those recommendations is above our job grade. But we'll provide that information." In other words, losing skilled NASA technicians could come back to bite the agency when it needed people to get the new Orion capsule ready to go.

All of a sudden, workers might matter.

Aside from the delicate issue of layoffs, the Augustine Commission was in Florida for a status report on NASA's favored design for a replacement spacecraft for the winged Space Shuttle. Since 2005, the agency had pushed the idea of the gumdrop-shaped Orion capsule and its Ares-1 solid fuel booster. Kennedy Space Center managers had tangible things to show the panel about how Ares was doing.

In March 2009, parts for the first test flight of the Orion capsule had been gathered in the Vehicle Assembly Building at NASA's launch site in Florida. The prototype rocket, known as Ares 1-X, was made from a leftover Space Shuttle solid fuel booster with a mock-up of Orion bolted on top. The trip would be a modest one. Mission managers planned to send the test rocket 150,000 feet into the air to see if a single solid fuel booster could successfully take off from a Shuttle launchpad. The booster would descend gently on parachutes, while the Orion capsule plopped unceremoniously in the Atlantic. While technicians tinkered with the test rocket, launchpad 39B underwent a noticeable facelift. The gantry was first used for the tragic launch of Space Shuttle *Challenger* in 1986, which ended in a deadly explosion. To accommodate the new Orion capsule, a trio of 600-foot lighting protection towers were erected around the pad, each resembling a tall, skinny oil derrick.

Despite the advantage of having a rocket on the pad to talk about, the Ares-1 rocket was under considerable criticism. Opponents were worried about excessive shaking that could damage Orion and possibly kill the astronauts. Another concern was "booster drift," where the ignited Ares rocket might move sideways and smash into the launch gantry. One additional scenario would be if the Ares meandered up and over the tower after liftoff, showering it with booster exhaust generating 6,000 degrees of heat. The builders of the new booster refuted all the critics, but the commission began looking at a number of different designs for the rocket that would carry the Orion capsule. Perhaps the most controversial came from a secret brotherhood of engineers.[15]

You Can Call Me "Atlas"

NASA has often been criticized for being a rigidly bureaucratic agency, and very resistant to change. Astronaut Andy Thomas once produced a video for the Internet Web site "YouTube" to illustrate the problem. The project was entitled "Barriers to Innovation and Inclusion" and featured a fictional low-level NASA engineer with an idea to improve a spacecraft. A volunteer from the agency portrayed the idealistic engineer who found her suggestion met with opposition and even veiled threats from other "actors" playing her supervisors and managers at NASA's contractors.[16]

Some engineers connected to the space agency worried it wasn't necessarily fictional.

The Augustine Commission eventually heard about an alternate spacecraft design, proposed by engineers with secret code names like "Atlas," "Delta," "Agena," and others derived from NASA rockets. They remained in the shadows for fear of losing their jobs to vengeful NASA managers. Ross Tierney, on the other hand, spoke his mind.

Tierney's accent betrays his hometown of Portsmouth, England. The former computer consultant made his living by building museum-quality models of NASA rocket launch gantries valued in the hundreds of dollars. He was sitting at the kitchen table of his girlfriend's Cocoa Beach townhouse in 2005 when he started adding up the costs of NASA's Ares-1 rocket and its cargo-carrying big brother, the Ares-5. NASA liked what it saw, but Tierney didn't. "It seemed a little high," he said of the price tag.

The two new Ares rockets meant possibly expensive renovations to the infrastructure at the Kennedy Space Center to allow technicians to work on the updated spacecraft. In other words, if someone sells his or her two-story house and buys one with three stories, one of the unexpected extra costs is having to purchase a taller ladder to clean the rain gutters. The same concern applied at the Vehicle Assembly Building at the Kennedy Space Center, which would need completely new work platforms to accommodate the bigger Ares-5 cargo rocket. Since the Vehicle Assembly Building has been handling the Shuttle and its external fuel tanks and booster rockets for thirty years, why not base NASA's next-generation spacecraft on that existing technology?[17]

Tierney put his concerns on a NASA-related Internet chat room and got immediate results. "I got fourteen engineers coming back to me within twenty-four hours saying, 'This is what we've been trying to push,'" says Tierney. "But their managers didn't want to listen." That led to the creation of the Web site "Jupiter Direct 2.0." It proposed a Shuttle-derived vehicle to replace the existing Space Shuttle. The new design would look somewhat like the Shuttle on the launchpad, just without the winged space plane. A taller external fuel tank would be equipped with a cluster of Shuttle main engines at the bottom, with standard solid rocket boosters on either side. The Orion capsule would be perched on top. A variant of the Jupiter rocket could carry cargo, making the bigger and more expensive Ares-5 heavy lift vehicle unnecessary. "You could swap out the Ares-1 vehicle with the Jupiter launch vehicle," proposed Tierney, "and completely delete all the work necessary for the Ares-5. You could save yourself something on the order of $20 billion in one flat move." The idea was presented publicly to the Augustine Commission during its first meeting in Washington, D.C. "Who are you guys?" asked panel member and former astronaut Leroy Chiao incredulously. Still, an independent review of the Jupiter design was ordered, and a "Pandora's box" of concerns over the Ares program had been opened.[18]

The Ugly Truth

During the public proceedings of the Augustine Commission, its chairman dropped a stream of gentle hints in the press about what kind of message his group would craft for the White House. Norman Augustine's central point was the panel wouldn't "fall on its sword" by recommending politically unpopular solutions to fix NASA. It would provide a range of alternatives, but President Obama and Congress would have to make the tough choices. The toughest decision concerned money.

Think of NASA as a dinner buffet, and Congress has one standard-size plate to fill. Once the diner had selected a steak, baked potato, and green beans and his or her plate was full, any additional menu items would mean something had to be left off to make room. From NASA's perspective, that menu included the International Space Station, the waning Space Shuttle program, and the prospect of visiting the Moon, Mars, or side trips to asteroids, better known as Near Earth Objects, or NEOs. From the perspective of the Augustine Commission, Washington's eyes were clearly larger than its budgetary stomach.

The International Space Station was set to be abandoned and

destroyed in 2016, and some proponents wanted to keep it operating longer to justify the billions spent to construct the complex. Still others favored flying the Shuttle past its 2010 retirement date to safeguard NASA jobs and to close the "gap" until the first launch of the Orion capsule. Norman Augustine believed that any of these options, like extending the Shuttle and International Space Station, carried a hidden cost that might hurt NASA in the future. "You can frankly forget about Ares-5, and you can probably forget about Mars for many, many years," he said. "On the other hand, if you don't continue with the Shuttle, if you live with the gap, you can build the groundwork for an exciting space program in the future. It's a tough trade, not a trade that we have to make. The President will have to make that trade, but we'll provide him with the options to make that trade." Still, the commission didn't want to give up on sending astronauts to Mars. In fact, Augustine pledged that a journey to the red planet would figure prominently in the panel's recommendations to the president. But again, money would be the deciding factor. "You can't go 60 percent of the way to Mars if you only have 60 percent of the money," said Augustine. Ultimately, the panel gave the Obama administration a list of possibilities. The options included extending the life of the International Space Station and the Space Shuttle, sending people to the Moon or abandoning Earth's nearest neighbor, exploring asteroids, or going to Mars.[19]

The Shuttle Winds Down

In May 2009, one month before the Augustine Commission was named, there was a tangible sign the Space Shuttle was breathing its last gasp. Seven astronauts made NASA's final trip to repair the Hubble Space Telescope. It was a misty-eyed moment made possible by a battle that had raged within the space agency for six years. Following the loss of Space Shuttle *Columbia* in 2003, NASA administrator Sean O'Keefe decided that visiting Hubble one last time was too risky and canceled the mission.

NASA's Second Shuttle Disaster

Columbia was doomed almost at the very beginning of its final mission when foam insulation from its big external fuel tank tore loose and struck a vital heat shield. The damage that resulted allowed hot gases from the friction of reentry to pour through the breech and into the interior of *Columbia*'s left-hand wing. Engineers theorized the blowtorch effect tore through the wing's inner structure and out the main landing gear door in the Shuttle's belly. The damage included the loss of the vehicle's hydraulics, so *Columbia* couldn't maintain a smooth path through the atmosphere with its heavily shielded belly protecting the crew. Investigators believe the powerless spacecraft began to skid on its tail through reentry. The Shuttle's tail rudder, wings, and payload bay doors tore away. Finally, the nose of the vehicle, containing the crew cabin and the astronauts, snapped off.

A Word on Survivability

The astronaut corps was reassured with news that the next-generation Orion vehicle would have a launch escape system, something the Shuttle lacked. The long, needlelike rocket bolted to the top of the new capsule traced its lineage back to the early days of Project Mercury. In case of a midair emergency, the rocket would fire and pull the capsule and crew away for a watery splashdown in the Atlantic. The final investigation into the *Columbia* disaster in 2003 left disquieting questions on whether the astronauts had a remote chance of surviving the breakup of their vehicle. The NASA report didn't voice an opinion one way or the other, but it mentioned the breakup of an SR-71 spy plane in 1966.

The SR-71 was developed by a division of Lockheed Martin called the "Skunkworks," which created a number of cutting-edge flying machines for the military. The SR-71 was considered the most sophisticated airplane ever built. The covert vehicle was designed to snap photographs during reconnaissance missions for the intelligence community. Its full capabilities were kept classified, but after the plane was decommissioned in 1990, it was allowed to make one final flight in 1999. The trip was between Los Angeles and Washington, D.C., and the aircraft traveled that distance in sixty-four minutes. The average speed of the

SR-71 was believed to be three times the speed of sound—so fast that the aircraft stretched six inches due to the heat of friction with the atmosphere. The windows were also said to heat up to 600 degrees, hot enough to cook your dinner on the windshield, if you so chose.

The final report into the *Columbia* accident stated that an SR-71 with a two-man crew aboard suffered a catastrophic failure in 1966. One of the passengers died; however, the pilot survived. The plane's crew members wore heavy launch and entry suits similar to those used on the Space Shuttle. The man who lived had his helmet visor down and locked. His suit automatically pressurized as his airplane broke apart around him. He fell out and parachuted to the ground. NASA states such a scenario would likely not have saved the *Columbia* crew during the breakup of their vehicle. But it is interesting that the agency mentioned the SR-71 incident in any case.[20]

The plan to resume Shuttle flights following *Columbia* did include extra safeguards for the crew. The Canadian Space Agency built an extension for the robotic arms on each of the surviving Shuttles. Once in orbit, the astronauts would spend most of their second day in space scanning their spacecraft for launch-day damage. To do this, the robotic arm would be used to grasp the extension, called the "boom." The long periscope contained lasers and a camera, which would go through a preprogrammed path along the Shuttle's exterior to look for problems.[21]

The International Space Station was another safety feature. Post-*Columbia* Shuttles would be equipped with extra food, water, and supplies. If their spacecraft was damaged and considered unsafe for landing, the crew would retreat into the station in something referred to as "operation safe haven." They would wait there until a rescue Shuttle was dispatched. This standard operating procedure made a last Hubble repair trip too risky. The observatory operated at a higher altitude than the space station and in a different orbital path around Earth. A Shuttle crew visiting the telescope would lack the propellent to reach the safety of the orbiting outpost. Also, there was the threat of space junk, which was leftover debris from previous rocket launches. There was more orbiting trash where Hubble was stationed, ranging from bolts to spent rocket boosters. The possibility of a Shuttle being hit by anything made a visit to the observatory even more dangerous.[22]

The Pitch for One More Hubble Flight

Astronaut Scott Altman was one of those attending the 2005 Space Exploration Conference in Orlando, Florida, which was the birthplace of the Orion capsule. Aerospace contractors lined up to offer different plans for NASA's next generation of spacecraft to map out the agency's future. Still, Altman was preoccupied with the Hubble Space Telescope. The veteran Shuttle commander, who went by the nickname "Scooter," led the last crew to repair the $1.5 billion observatory. Along with that distinction, Altman was also known in aerospace circles as the fighter pilot who did the fancy flying that made actors Tom Cruise and Val Kilmer look good during the motion picture *Top Gun*. He defended his boss's decision against another Hubble mission. But when asked if he'd ride the Space Shuttle on one final repair flight to the aging observatory, Altman responded, "Sure."[23]

John Grunsfeld was thinking along those lines as well. He had spent more than half of his career as an astronaut making repairs to Hubble. He had seen it go from a political millstone around NASA's neck to a shining achievement for the agency. He wasn't going to throw in the white towel and surrender just yet. "Really it was a question of life extension," says Grunsfeld. "If we gave up on it now, Hubble would have maybe five more years of life left in it, and that would have given us until 2008." After a shaky start following its launch aboard Space Shuttle *Discovery* in 1990, Hubble was at the top of its game, scientifically speaking. Grunsfeld's two previous servicing missions had installed new optical instruments, including the Cosmic Origins Spectrograph. It was designed to detect the basic chemical building blocks of the universe, and at twenty times the strength of any previous device.

NASA was still recovering from the *Columbia* disaster, and top management remained adamantly against sending astronauts to add the batteries and gyroscopes to keep Hubble healthy. Grunsfeld and the telescope's supporters kept busy with alternatives. One was careful management of the parts already aboard Hubble to keep it running as smoothly as possible. The other was to design a robotic mission that could swap vital parts without risking human life. "All of that," says Grunsfeld, "allowed the team to keep ready for a future Shuttle servicing mission even though there wasn't one on the flight schedule."

The successful launch of Space Shuttle *Discovery* in the summer of 2005 put all of the new post-*Columbia* safeguards through successful tests in Earth orbit. That, plus the presence of a new NASA administrator, changed Hubble's fortunes. Michael Griffin was more receptive to one final servicing mission. In October 2006, both John Grunsfeld and Scott Altman were named to STS-125, which would visit the telescope one last time. It would be one of many good-byes as the Shuttle program began to wind down.[24]

Hubble's Ups and Downs

The problems associated with the Hubble Space Telescope began in the 1970s when the PerkinElmer company built the eight-foot-wide main mirror. It would act as the observatory's primary "eye" in space. The smooth surface was carefully, but improperly, ground. When Hubble was left in orbit by a Space Shuttle crew in 1990 and its optics were switched on, the result was a useless blur. Hubble was, at that point, little better than a ten-ton paperweight orbiting the Earth. Usually, following successful space missions, the first press conference is crowded with high-level NASA managers eager to share the news with the world. When the agency admitted the problems with Hubble, Ed Weiler was on his own. He was chief scientist on the orbiting observatory, and his job was to report news no one wanted to hear. "We're all frustrated, obviously," said Weiler. The flaw in the mirror created what was called a "spherical aberration." Simply put, when Hubble looked at something, there was a blurry glob in the field of view. "We can still do important and unique science until the problem is resolved," he insisted.[25] Not everyone was convinced. U.S. senator Barbara Mikulski, who chaired a Senate committee that helped control NASA's budget, declared Hubble a $1.5 billion "techno-turkey."[26]

The First Fix-It Mission

In 1993, astronauts Dick Covey, Ken Bowersox, Kathryn Thornton, Story Musgrave, Jeffrey Hoffman, Tom Akers, and Claude Nicollier blasted off aboard Space Shuttle *Endeavour* with a tough job ahead. They had to restore the observatory's blurry vision and try to erase the

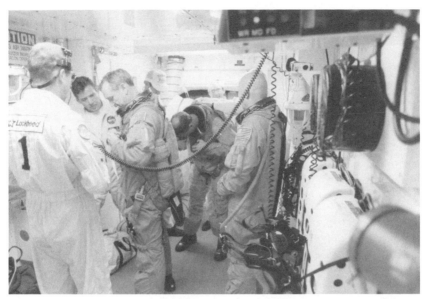

Astronauts Tom Akers and Story Musgrave prepare to board Space Shuttle *Endeavour* for the mission to restore the vision of the Hubble Space Telescope. Courtesy of NASA.

blemish it left on NASA's public image. Spacewalkers had never touched the telescope while circling the Earth at 17,000 miles per hour. This mission would prove that on-orbit maintenance was possible with a record five spacewalks. The astronauts installed a box designed to compensate for the misshapen main mirror that left Hubble largely blind. The unit was called the Corrective Optics Space Telescope Axial Replacement, or COSTAR. NASA hoped that its tiny pick-off mirrors, some the size of a nickel, would straighten out the observatory's fuzzy view like a pair of eyeglasses on a person.[27]

The *Endeavour* crew did its work and landed successfully on December 13, 1993. One agonizing month later, the press was summoned for a briefing on Hubble. Ed Weiler, who had shared the disappointing news of the telescope's failure in 1990 all by himself, was on the stage once again. But this time he had company. On January 13, 1994, there was good news to share, so NASA administrator Dan Goldin and Senator Mikulski took the center chairs to address the reporters. The lawmaker from Maryland held up two photos taken by Hubble's Faint Object Camera. One picture, which was taken before *Endeavour*'s repair mission, was fuzzy. The "after" photo was sharp. "The trouble with

Astronaut Dick Covey suits up for the mission of Space Shuttle *Endeavour* to make the first servicing flight to the Hubble Space Telescope. Spacewalkers installed corrective lenses to fix the observatory's blurred vision. Courtesy of NASA.

Hubble is over!" Mikulski exclaimed.[28] NASA breathed a collective sigh of relief.

Getting Acquainted with Hubble

John Grunsfeld had spent more of his career tinkering with Hubble than had any other astronaut. His first two space flights, however, had nothing to do with the observatory. He first flew aboard *Endeavour* in 1995 to aim telescopes aboard the Shuttle at sources of ultraviolet light. His next mission was a visit to the Russian space station *Mir* in 1997. National Public Radio (NPR) listeners may remember that flight because of an unusual appearance Grunsfeld made on NPR's *Car Talk*, a call-in program where listeners could get advice on problems with their cars and trucks from hosts Tom and Ray Magliozzi.

Grunsfeld was schooled at Massachusetts Institute of Technology, as were the two *Car Talk* hosts, who went by the nicknames "Click and Clack, the tappet brothers." They also ran an auto garage in Boston. Years later, Tom and Ray were fixtures on public radio, and Grunsfeld was on his way to the *Mir* space station. The astronaut and the radio program's producers had concocted a plan to stage a prank phone call from orbit.[29] "That was the only time I was ever nervous in space," recalled Grunsfeld. "They played the previous call over the communication line, so I heard Tom and Ray's response. I knew I had to come up with something." Grunsfeld identified himself as "John from Houston." He described over the phone how he occasionally rode in a government-owned vehicle (the Space Shuttle), and how it ran really rough during the first two minutes (during the early part of the blastoff to orbit). Tom and Ray apparently didn't suspect anything, until Grunsfeld reminded them how they maintained his car in the late 1970s. Then the jig was up.

Grunsfeld's next trip into space was in 1999. It was his first visit to Hubble. The list of tasks on the mission included four spacewalks to install $69 million worth of new equipment, including new gyroscopes and a new main computer. Grunsfeld was joined in orbit by crewmates Curt Brown, Scott Kelly, Mike Foale, Steve Smith, Jean-Francois Clervoy, and Claude Nicollier. All of the veteran astronauts huddled around the windows in *Discovery*'s flight deck during the delicate rendezvous.[30] "The first impression you got was that it was really big," Grunsfeld recalled. "It starts off as a little star in the distance. Then everybody gets out the binoculars, and you can see the solar panels." By the time the astronauts reached out with the robotic arm to grasp the telescope and lock it in *Discovery*'s cargo bay, Hubble appeared full size, nearly as big as the Shuttle itself.

Then it was time for Grunsfeld's first spacewalk. He was teamed with Steve Smith to replace the gyroscopes that would help Hubble point at targets in space. To do the work, each man would be dressed in a 300-pound spacesuit with a helmet and a fishbowl-type visor. Grunsfeld would clamp the boots of his suit to the end of the Shuttle's robot arm so European astronaut Jean-Francois Clervoy could move him around like a utility worker perched on a cherry picker crane. "I was about three feet away from Hubble," Grunsfeld recalled, "and I just had to reach out and touch it to make sure it was real. It was covered with this shiny aluminized skin so it reflected the Earth and everything around it. It was a work of art." Hubble's exterior wasn't blemish-free after nine years in orbit. The solar panels were slightly bent, and there were small dents from meteoroids that had peppered its hull. But once the hinged doors on the telescope's main body were swung open, its interior was even more surprising. "Once you opened it up," says Grunsfeld, "the inside had been in the complete vacuum of space for all that time. It was pristine in there."

After Smith and Grunsfeld did their spacewalk, the crew took a day off, and then Foale and Nicollier took their turn on spacewalk number two. Nicollier had the advantage of having visited Hubble before, in 1993. Foale had the experience of knowing firsthand how badly a spaceflight could go. He was aboard the Russian space station *Mir* in 1997 when his two cosmonaut crewmates were guiding a robotic cargo craft, called a progress vehicle, toward docking. The vehicle went out of control and headed toward a collision with the orbiting laboratory. The Russians called Foale over the radio to get their Soyuz escape vehicle ready for an immediate evacuation.[31] "I was floating between compartments toward the Soyuz," says Foale, "when there was a thud, and the whole space station moved to the left around me." The thud was followed by a strange sensation in his ears as the air inside *Mir* began leaking out. The damaged compartment on the station was sealed off, and he and his crewmates survived the ordeal.

Still, the experience aboard *Mir* didn't cure Foale of nervousness when it was his turn to tinker with Hubble. He and Nicollier would change out the main computer. "It was nerve-racking," recalled Foale. The telescope might be able to operate with a misaligned gyroscope or battery, but without the computer, Hubble was dead, and the spacewalkers knew it. "My little hands could have killed Hubble forever," says Foale. "It was a horrible responsibility, which I took and enjoyed having accomplished. But at the time, I was very worried about it." Nervousness around the billion-dollar telescope was something with which John Grunsfeld could identify. Working in heavy spacewalking suits was like handling fine china while wearing thick mittens. "The mantra I instilled in the crew was 'don't break the telescope,'" he says. "There are a lot of delicate things in there." Adding to the complexity of the work that needed to be done was that Hubble needed be switched off completely for part of the mission. That had never been done before. Even performing every task perfectly would be valueless if the space-craft didn't come back to life, which it did.[32]

John Grunsfeld would return to Hubble in 2002 with Scott Altman, Michael Massamino, Duane Carey, Nancy Currie, Rick Linnehan, and James Newman. It was a diverse crew. Linnehan was a veterinarian, and Newman and Currie had snapped together the first two pieces of the International Space Station in 1998. "The solar panels were a little more bent, and there were more tattered edges to the protective covers," recalled Grunsfeld upon seeing the telescope again. "It looked better when we deployed it with spiffy new solar arrays."

The astronauts would also install Hubble's first new instrument in five years. The Advanced Camera for Surveys was designed to photograph big patches of the universe instead of zeroing in on one target. It would later fail, along with another device called the Space Telescope Imaging Spectrograph. Those parts were on the repair list for the final repair mission to Hubble, and the work would again stretch the limits compared to the kind of jobs that had been done previously.[33] Just as on his first visit to the telescope, Grunsfeld admitted to an emotional connection with the observatory. "When I saw it the last time," he recalled of the 2002 flight, "I was happy to see it in space. It was thrilling when we deployed it as a working telescope again. But there was this sadness that I might not see it again."

The Last Fix-It Mission

Grunsfeld would get one more chance, but only after troubles and tragedy. The year after his last servicing flight ended, Space Shuttle *Columbia* and its crew were lost in the skies over eastern Texas. One year later, Hubble started falling on hard times. The Space Telescope Imaging Spectrograph stopped working when its power supply failed. In 2006, the final repair mission to the telescope was authorized. Grunsfeld would be accompanied by mission commander Scott Altman, pilot Gregory Johnson, robot arm operator Megan McArthur, and spacewalkers Mike Massamino, Andrew Feustel, and Michael Good.[34]

Astronauts John Grunsfeld and Andrew Feustel tinker with the Hubble Space Telescope. During the work session, the spacewalkers replaced batteries on the observatory, along with a fine guidance sensor and protective blankets on the telescope. Photo courtesy of NASA.

The next year, Hubble's Advanced Camera for Surveys failed due to a short circuit. The work list for STS-125 was getting longer.

A Full Agenda for the Repair Crew

Each servicing mission to Hubble built on the previous flight. Follow-up missions to the observatory became more and more complex as NASA's spacewalking astronauts became more familiar and confident with tinkering on Hubble. By the time John Grunsfeld and the final servicing crew were assigned, even more work was expected.

STS-125, as the repair flight was known, would install new batteries and gyroscopes as well as a new camera the size of a baby grand piano. This was common for Hubble flights. The telescope was built with doors that swung open and big components that slid in and out on rails. Spacewalkers, on comparatively easy tasks, might be expected only to plug in power cables to accomplish their work. The last Hubble mission included actual repairs to two instruments. Grunsfeld and Mike

Massamino removed dozens of screws from the Advanced Camera for Surveys and from the Space Telescope Imaging Spectrograph. Once the boxes were open, Massamino would try to install a new power supply on the Space Telescope Imaging Spectrograph. Grunsfeld's challenge was to insert new transistor cards into the Advanced Camera for Surveys while wearing heavy spacesuit gloves.[35] "I practiced taking out the screws with my eyes open and closed," said the astronaut. "I did all the 'Karate Kid' techniques to make it as much of a Zen experience as I could. I would say to myself, 'I am the screws, I am the circuit boards.'" Grunsfeld also kept a spare plate with screws he could practice with while sitting at his desk.

More Troubles for Hubble

Grunsfeld and his crewmates spent months training for their mission. That included hours with the spacewalkers in their suits floating in a deep swimming pool to simulate the weightless environment in which they work in orbit. The astronauts visited the Kennedy Space Center in August 2008 for the traditional countdown dress rehearsal at the launchpad. For the four rookies on the crew, it would be the first time they had actually sat inside the Shuttle on the pad. The practice wrapped up on September 24. Six days later, the astronauts' job got harder.[36]

Grunsfeld was at home when he got a phone call from a friend at the Goddard Space Flight Center in Maryland. The facility acted as the control center for Hubble. The news wasn't encouraging. The telescope's data transmitter, which had operated just fine since 1990, stopped working. "I was playing the denial card," says Grunsfeld. "Things on Hubble go off-line all the time. I figured the people at Goddard were smart, and it would be a short scare and we'd be OK."

It didn't turn out that way. Mission controllers had to take the risky step of firing up a back-up communication channel inside the telescope, which hadn't been used in eighteen years. That idea worked, but then the main computer and a camera failed to boot-up correctly. Grunsfeld's mission was officially postponed until at least April 2009. The crew got the news during a practice session at the Johnson Space Center in Houston.[37] "I was surprised by how fast the decision to delay was made," says Grunsfeld. "We were just days from going into quarantine

Spacewalker Mike Massamino grins for the camera between jobs on the Hubble Space Telescope, while crewmate Michael Good works on the observatory. The two astronauts performed repairs on the Space Telescope Imaging Spectrograph during NASA's final repair flight to the Hubble. Photo courtesy of NASA.

and weeks before liftoff. Usually the science and Shuttle people take their time in making this kind of decision." This time they didn't. As big a disappointment as the launch delay was for the crew, there was general agreement that it could have been a lot worse. The transmitter could have failed after the astronauts had finished their mission and landed back on Earth. "Now that would have been a disappointment, to use an understatement," says Grunsfeld. "I'm glad we didn't do all that work, only to have no way to fix the transmitter afterwards." The spacewalkers would return to the practice pool to learn how to replace the transmitter along with everything else on their list. If the mission succeeded, Hubble could operate until at least 2013, perhaps longer. If not, its swan song could come much sooner.

Travel Plans for 2020

No matter what happened to Hubble during the final repair flight, at least the astronauts were confident that they didn't face the risk of killing the telescope as crews had on previous missions. The Advanced Camera for Surveys and the Space Telescope Imaging Spectrograph weren't working before the liftoff. Even if the repairs weren't successful, at least Hubble would be no worse off. The concerns were unfounded. The ambitious repair flight, with the groundbreaking repairs, restored Hubble's instruments and most of its lost capabilities.

By 2020, NASA astronauts may be back on the Moon. The International Space Station may or may not still be operating. The Space Shuttles will have long since been relegated to museums, so there will be no chance of plucking the Hubble Space Telescope from orbit for a safe return to Earth. The new Orion capsule can't bring back large pieces of hardware from space. Instead, a robotic mission is being considered to clamp a cluster of booster rockets to the observatory. The point is to push Hubble down harmlessly in the atmosphere to burn up and plunge into the Pacific Ocean.[38]

Astronaut John Grunsfeld may be on a cruise in the Pacific at that time. He plans to be with colleagues who worked on the telescope for the observatory's swan song, whenever it might be. "We'll get on a cruise ship and watch it reenter," he says. "It'll be a celebration, because by any measure, the telescope has exceeded anyone's expectations of what it can do, what it's discovered, and the value it's brought to everyone."

While Grunsfeld and his colleagues watch for the fiery streaks across the sky as Hubble burns up, NASA plans to be well on its way to sending humans to Mars. It will involve the dangers and challenges of pioneering the Old West that one astronaut is familiar with—not because of his own personal experience, but that of a distant ancestor.

pioneers past
and present

If anyone understands the comparison of a manned mission to Mars to taming of the American West, it's Jerry Ross. He's one of the Space Shuttle astronauts preparing to "hang up his spurs" at the end of the program. The seven-time Shuttle crew member says he's ready to pass the baton to the next generation of astronauts, whose job it will be to launch NASA's replacement for the Shuttle, perhaps return to the Moon, and go on to Mars. The retired air force colonel may also have some observations to offer the "new kids" on how the challenges of establishing a human presence on Mars might compare to life in the American West in the late nineteenth century.

Ross found that space travel and life in the Old West run in his family. The discovery came from his interest in genealogy, which he pursued while preparing for one of his final missions in orbit. The journey took him from modern-day Houston, Texas, to the doorstep of Miners' Delight, Wyoming, in 1870. Just as Ross helped to further space exploration, the brother of his great-great-grandfather had worked to tame the American frontier.[1]

But making this connection between pioneers past and present would only come after some hard work in the Space Shuttle astronaut corps.

Ross began his career at NASA after he watched the Apollo Moon program breathe its last gasp in the mid-1970s. America had already finished the effort to put astronauts on the lunar surface. The last three scheduled missions were canceled due to budget cuts in Washington. Those rockets and capsules would be cannibalized and used for the four final flights related to Apollo. Nine astronauts would fly aboard the Skylab space station in the early 1970s. The orbital workshop was built from the hollowed-out upper section of the Saturn V rocket that was supposed to carry people to the Moon on Apollo 20.[2] After the three missions of Skylab came the Apollo-Soyuz Test Project. That mission featured docking NASA's last Apollo capsule with a Russian Soyuz vehicle. It would team up Mercury astronaut Deke Slayton with cosmonaut Alexei Leonov, who performed the world's first spacewalk in 1965.

After that, Apollo was dead, and Jerry Ross's generation got its chance. The first group of Space Shuttle astronauts was named in 1978, three years before the first launch of the new winged spacecraft. That first collection of trainees included people like Hoot Gibson, who would later dock the Space Shuttle *Atlantis* to the Russian space station Mir in 1995. His classmate Norman Thagard would be waiting aboard the orbital laboratory as the first American to live and work there.[3]

Jerry Ross would be part of the second group of astronauts, who were selected in 1980.

He would spend the next five years working behind the scenes in support of the first Shuttle crews to go into space. His first mission in orbit came in 1985. *Atlantis* would carry seven astronauts and a cargo of commercial satellites for Australia and Mexico.[4] Better yet, Ross's job included his first spacewalk, along with his crewmate Sherwood Spring. The two men ventured outside during the *Atlantis* mission to practice assembling long, girderlike structures that might be used during the construction of a future space station. At that time, NASA was more than a decade away from building the International Space Station, so Ross's session outside in a 300-pound spacesuit was just practice. All of the spacewalkers breathed pure oxygen through a mask prior to suiting up to purge nitrogen from their blood. The point was avoiding nitrogen bubbles in their bloodstreams in the lower air pressure of their spacesuits. Scuba divers refer to this condition as "the bends" due to how painful it can get.[5] Then came the heavy spacesuits, which astronauts put on in sections. Ross and Spring slid on the heavy pants, and then clamped on the upper part, corresponding to a shirt. The ensemble was completed by locking on gloves and the big fishbowl helmet. Both men, fully outfitted, then squeezed inside *Atlantis*'s airlock and pumped out the air so they could float outside in the vacuum of space.

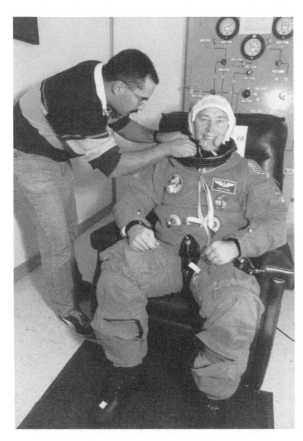

Astronaut Jerry Ross suits up for liftoff. This particular shot was taken prior to the launch of *Atlantis* on STS-74, the second docking mission with the Space Shuttle and the Russian space station *Mir*. Courtesy of NASA.

"I recall sticking my head out of the airlock for the first time," says Ross. "I wanted to give out a 'war whoop' of happiness, but I was worried that someone in Houston would think 'Ross just lost it' and call me back in. So I kept quiet."

That flight put Ross on a path toward more missions and more spacewalks. It also gave him experience that future rookie spacewalkers would seek out. That would include not only how to work in space but also how to make the most of the experience on a personal level, such as his advice to always remember what they see. Spacewalkers see the world from space through a wide visor stretching from ear to ear while floating outside. It's a better view than what astronauts on the Shuttle and space station see through the triple-paned windows of either spacecraft. Ross recommends that each new spacewalker select an object they can see from their vantage point 250 miles above Earth, and to burn that image into their memories. It could be a city lit up at night while the Shuttle was on the side of Earth shielded from the Sun, or a mountain range, or an island. Ross has his favorites as well.[6]

"For me, it's the sunrises and sunsets," he says. "They are incredible, and they happen very, very quickly compared to what you get on the ground. We're moving at five miles per second, and the colors are amazing. We've never been able to capture it with cameras and film."

There's much less to say about Ross's second mission in orbit. The crew deployed a secret satellite for the Pentagon.[7] Ross's third flight was considerably less high-tech. In 1991, he joined four astronauts with the job of delivering the seventeen-ton Gamma Ray Observatory to orbit. Ross and rookie spacewalker Jay Apt had no specific job to do regarding the boxy telescope, nicknamed GRO. Their spacewalks included more practice for future space station work, like testing a way for crew members working outside to slide from place to place to perform various tasks. However, they stood by in their heavy spacesuits in case something went wrong during the deployment of the observatory clamped into the cargo bay of Space Shuttle *Atlantis*. The rest of the Shuttle crew included first-time Commander Steve Nagel and rookie pilot Ken Cameron. Astronaut Linda Godwin would operate the Shuttle's fifty-foot-long robotic arm and deploy the satellite by remote control. Unloading the observatory meant first extending its main dish antenna on the end of a long mast. It would be tucked up against the body of the telescope for the trip to space, and then swung out to its full length. After that, its twin solar panels would be unfolded to provide electricity for the heaters that would protect the telescope's delicate optics. Any problems in the sequence could mean canceling the release and the space mission ending in failure.[8]

Early in the deployment process, the antenna arm became stuck. The challenge for the Shuttle crew was that there was no clear indication of what was wrong or how to fix it. Ross and Apt would have to improvise a repair plan while floating on what would be considered NASA's first unscheduled contingency spacewalk.[9]

Apt and Ross headed to the lower deck of the Shuttle's two-story crew compartment to dress in their heavy white spacesuits. After checking their helmets for an airtight seal, the two entered *Atlantis*'s airlock, closed the hatch, and pumped out the air. There wouldn't be far to go

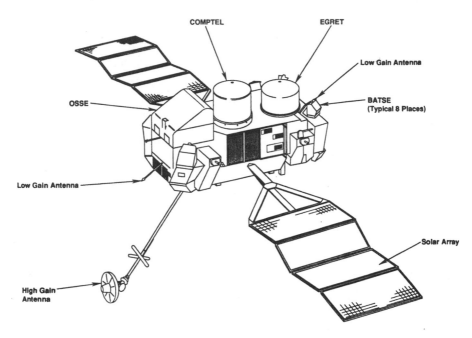

COMPTEL EGRET

Low Gain Antenna

OSSE

BATSE
(Typical 8 Places)

Low Gain Antenna

Solar Array

High Gain
Antenna

Gamma-Ray Observatory (GRO)

The Compton Gamma Ray Observatory was carried to Earth orbit by Space Shuttle *Atlantis* in 1991. It was the second of NASA's "great observatories" including the Hubble Space Telescope and the Chandra X-Ray Observatory. Its deployment almost failed when its main antenna became stuck shut. Courtesy of NASA.

to reach the worksite since the Gamma Ray Observatory nearly filled the entire cargo bay. Ross and Apt's crewmates kept watch through two windows at the rear of the crew cabin that overlooked the spacewalkers and the ailing telescope.

"The worst part was we didn't know what was wrong," says Ross. "We were trained how to do back-up tasks, but how the antenna hung up and whether we could deploy it was a big unknown when we went out the hatch."

There would be no war whoop of happiness from Ross on this space-walk either. The fate a $2 billion telescope hung in the balance. The Gamma Ray Observatory was one of NASA's four "great observatories," in the same league as the Hubble Space Telescope or the Chandra X-ray Observatory.[10]

"It reminded me structurally of a steam locomotive," recalled Ross. "It had these big structural pieces. It was sturdily built and pretty much filled up the whole payload bay."

Ross took hold of the antenna boom and shook it several times, and it finally popped loose. The Gamma Ray Observatory was deployed as planned. It studied gamma ray bursts from exploding stars for nine years before its mission was concluded. Mission managers then guided the satellite back into Earth's atmosphere to burn up harmlessly.[11]

Ross returned from this space mission a hero. He flew again in 1993 aboard *Columbia* on a science mission with experiments from Germany, and then in 1995 on a docking trip between the Space Shuttle and the Russian space station Mir.

Ross was next assigned to the Shuttle crew that would begin construction on the International Space Station. *Endeavour* would carry the drum-shaped *Unity* module and snap it to the school bus–size *Zarya*, a compartment built and launched by the Russian space program. The job of NASA's first chunk of the orbiting complex would be to connect *Zarya* to other American-built sections, such as a science laboratory called *Destiny*, and compartments being designed by the Europeans and Japanese. *Zarya*'s function was to act as a communications and power center, and to control the position of the station with jet thrusters. However, Space Shuttle *Endeavour* couldn't blast off until a Russian Proton rocket successfully carried *Zarya* to the right spot in Earth orbit. The liftoff from a launchpad in the former Soviet nation of Kazakhstan encountered delays, and Jerry Ross and his crewmates had to wait.[12]

To occupy some of that spare time, he took a three-week vacation and headed to his hometown of Crown Point, Indiana. One reason was to visit his family, but he also wanted to dig into his past.

Crown Point is in northwest Indiana, about thirty miles from Chicago. Aside from Ross's exploits in space, the town's greatest claim to fame was a jailbreak by John Dillinger in 1934. The Indianapolis native was under arrest for allegedly killing a police officer during a bank robbery. He escaped, much to the embarrassment of local police, by carving a fake gun out of either a bar of soap or a piece of wood.[13]

Jerry Ross's family had lived there for more than a century, and his wife's family had similar roots in the nearby town of Sheridan. From

Astronaut Jerry Ross's ancestor Daniel Dillabaugh and his family, circa 1890, following his work to help pioneer post–Civil War Wyoming. Photo courtesy of Jerry Ross.

the perspective of genealogy, that made Ross's search for his ancestors a lot easier. He spent a week and a half at his mother's home and an equal amount of time with his in-laws.

"I raided all of my relatives' cellars and attics," says Ross. "I visited every library I could find, tromped through cemeteries, and went to courthouses. Anywhere I could find details."

The search led Ross to Daniel Dillabaugh.

There were references connecting this man to battles during the U.S. Civil War, the postwar cavalry, and gold mining in Wyoming in 1870. But even more details came to light after Ross tracked down letters written by Dillabaugh that were part of the collection at the Michigan State Library. Later, a diary was found that had been written by his wife, whom he had married in the 1880s. The book had been kept by the stepdaughter of one of Dillabaugh's sons. This research revealed an unknown branch of Ross's family tree on his mother's side. Her maiden name was Dillabaugh, and Daniel was Jerry Ross's great-great uncle. He even found an old photograph of what his illustrious ancestor looked like.[14]

Daniel Dillabaugh was relatively short and not too heavily built. The native of Canada moved to the United States and fought for the Union during the Civil War. That included campaigns like the second battle of

Bull Run and Chantilly. Dillabaugh and his fellow troops faced forces led by General Robert E. Lee in both fights. Jerry Ross's predecessor was wounded in the shoulder during Chantilly, leaving him partially paralyzed.[15]

"He did an amazing amount of things," says Ross. "He cut down trees, chopped wood, and farmed, even though the Civil War records showed he was partly paralyzed in that arm."

After the Civil War, Dillabaugh joined the U.S. cavalry in Wyoming to help settle the territory and keep peace with the Indians who lived there. He later turned in his blue uniform for a pick and shovel to prospect for gold in the tiny outpost of Miners' Delight. His name first appeared in census records of 1870, with his occupation listed simply as "miner." He was one of eighty residents living in a remote bastion of civilization.[16]

Astronauts settling on Mars in the 2030 time frame might find similarities to the challenges that Daniel Dillabaugh faced in the Wyoming territory in the 1870s. NASA may have greater technology, but the

This is one of the few surviving structures from the late-nineteenth-century gold rush at Miners' Delight, Wyoming. Pioneers like Civil War–veteran Daniel Dillabaugh faced isolation, difficulty in receiving fresh supplies, and hostile environmental conditions— challenges similar to what twenty-first-century astronauts will likely endure on the surface of Mars. Photo courtesy of Jerry Ross.

agency will send people on a greater challenge. Their destination will seem, in many ways, as alien and hostile as what Dillabaugh encountered on Earth in the late nineteenth century.

Jerry Ross wasn't the only person to delve into the history of Miners' Delight and Daniel Dillabaugh. Eight miles to the south sits the larger community of South Pass City, Wyoming. It's another mining town, which has been carefully preserved by the Wyoming Department of State Parks and Historic Sites. State Curator John Lane spent years going through the records, diaries, census forms, and shop ledgers of Dillabaugh's time in Miners' Delight.

Life in that remote spot in Wyoming was one of isolation, something members of a future base on Mars will face.

"The outpost was about 100 miles from the nearest railroad at that time," says Lane of Miners' Delight. "This is significant because of the difficulty of traveling back and forth. During the winter, it was especially tough and done under only the most dire of circumstances."

After leaving the relative comfort of their railroad cars, residents of Miners' Delight faced the final leg of their journey on horseback or on foot to reach the outpost, with their belongings and supplies loaded onto wagons or pack animals. It could take a week or more to finish the trip. A broken tool, a medical emergency, or a shortage of provisions could mean severe hardship or worse, since help was a significant distance away.[17]

Likewise, NASA's first settlers on Mars will face even more challenging supply routes. A manned rescue spacecraft or even a robotic supply vehicle might take six months to arrive from Earth, assuming the two planets are close together in their orbits around the Sun. Only then can rockets easily travel between Earth and Mars.

An unwelcoming climate was another problem faced by settlers in post–Civil War Wyoming, which will also be encountered by future astronauts on Mars.

Miners' Delight is located in the rolling foothills of the southern edge of the Wind River Mountains. The terrain is at an elevation of 8,000 feet. That meant sparse vegetation and no trees large enough to provide timber to construct homes or other buildings. Daniel Dillabaugh and his neighbors traveled ten to fifteen miles to find enough wood to make suitable shelters.[18] NASA crews heading to Mars will have to bring much of what they will need to survive on the planet. Water might be extracted from ice at the polar caps, but only after considerable effort.[19]

In Wyoming, winter lasts for six brutal months. Even including the height of summer, there are only two months out of the year with no frost that would kill crops. Settlers had to plan carefully and carry in supplies to ensure there would be enough to eat, especially when eight-foot snowdrifts cut off Miners' Delight from the rest of the world. No snowdrifts are expected on Mars, but that planet has its own weather patterns that can kill. Astronauts will have to watch for electromagnetic storms caused by solar flares, which could expose them to deadly radiation and fry the electronics in their equipment.[20]

It was the promise of gold that drew people like Daniel Dillabaugh to Miners' Delight. The riches were short-lived. About 60,000 ounces of gold were taken from the central mine, as well as smaller operations, during the early 1870s. The veins eventually dried up, and the people moved on. The Wyoming Territory was known more as a place people passed through while going somewhere else. Those who live in the "Equality State" say things haven't changed much over the decades since then.[21] Astronaut Jerry Ross is constantly reminded of what settlers of the American West went through in order to pave the way for the modern age.

"Every time I fly over that part of the West," says Ross, "and see those vast miles, and the territory they had to go through and the technology with which they had to do it, it's easy to see why many of them didn't survive very long."

NASA astronauts, he insists, will have an easier time on Mars because of greater technology. Of course, that's future technology that the agency hasn't created. That's why the U.S. space program may return people to the Moon first. NASA needs to prove that humans can live there before venturing on to the more remote location of Mars.

the moon,
one baby step

A manned trip to Mars will present dangers and challenges never before faced by NASA astronauts. Crews there will be required to "live off the land" by harvesting resources from the planet and coping with conditions there. These were hardships familiar to the men and women who sought new lives in the American West.

Perhaps the best illustration of how daunting a trip to Mars might be was the last visit to the Moon. That was 1972, when astronauts Gene Cernan and Harrison Schmitt spent three days exploring the lunar region known as Taurus Littrow. A fourth day outside might have been too dangerous. The problem was Moon dust, which had caked into the joints of their suits where the crewmen's helmets, gloves, and other suit parts snapped on. The astronauts also carried lunar dust inside their landing craft, where some Apollo crews reported eye irritation and a smell similar to gunpowder.[1] If NASA ever hopes to establish a manned base on the surface of Mars, the agency admits it may have to conquer the Moon first.

The challenges of Apollo 17 were for a lunar visit lasting for three days. By comparison, crews will spend months on Mars and perform dozens of spacewalks on the red rocky surface. Those astronauts and their gear will have to hold up under relentless conditions without help from Earth. If a life-threatening emergency occurs on Mars, a quick return trip may not be possible. This would be especially so if Earth was too far away in its orbit around the Sun. Just like the pioneers who settled the American West, astronauts on Mars will have to survive by their wits and the tools they have with them. That's why some at NASA want to return to the Moon before venturing to Mars. The agency needs to gain confidence in the technology and skills that will be used on the red planet. The Moon could be a good place to practice.

Reality Falls Short of Dreams

Moon bases where people would live and work are still the stuff of science fiction, and outposts on Mars are even farther out of reach. Perhaps the most famous Moon complex is the one depicted in the motion picture *2001: A Space Odyssey*. Science fiction writer and futurist Arthur C. Clarke envisioned a palatial base complete with a dome-shaped hangar with doors that retracted like the wedges of a pie to receive incoming spacecraft. The creator of iconic fictional characters like "Hal" the computer often brainstormed concepts for technology now taken for granted. NASA also found itself measured against the ideas Clarke put down on paper. The agency's accomplishments were far from the lavish ideas the writer had in mind. That was how Clarke felt during his final visit to the Kennedy Space Center.[2]

Space Shuttle launches routinely attract celebrities. U.S. presidents and other heads of state have occupied seats in the grandstands or the launch control center. Even TV personalities have shown up at the press site to talk with reporters on launch day, ranging from undersea explorer Jacques Cousteau to actress June Lockhart, who played Maureen Robinson in the campy television series *Lost in Space*.[3] Having luminaries around was routine for the media covering NASA, but August 18, 1994, was different. Space Shuttle *Endeavour* was on the launchpad, and Arthur C. Clarke was in the crowd.

The writer's previous visits to the Kennedy Space Center on launch day were for Apollo 11 in 1969, Apollo 12, and Apollo 15.[4] In 1994, the Space Shuttle was poised for the second Space Radar Laboratory mission. A forty-foot-long, thirteen-foot-wide flat radar antenna filled the spacecraft's cargo bay. Astronauts Mike Baker, Terry Wilcutt, Jeff Wisoff, Dan Bursch, Steve Smith, and Tom Jones would use the system to study Earth. The list of targets included the oceans and the movement of landmasses. Scientists even wanted to use the radar to pierce the heavy cloud cover of Rwanda to observe the mountain gorillas that lived there. That was of particular interest to Clarke.[5]

Of course, not even the astronauts were guaranteed a routine liftoff. Many Shuttle launch attempts end in a scrub, and that was the case for *Endeavour*. The astronauts and their audience sat out the countdown, and everything looked good until the very end.

Blastoff for the Space Shuttle comes in two stages. First, the three liquid-fueled main engines beneath the spacecraft's tail rudder ignite at six seconds before takeoff. The motors roar to life about a millisecond apart from each other to minimize the shock wave bouncing back against the vehicle. Then the solid fuel booster rockets fire, but *Endeavour* didn't get that far. The craft's onboard computers sensed that the turbo pump on the third main engine was running too hot. The system aborted the launch automatically and shut everything down just two seconds before takeoff.[6]

Once a Shuttle's engines ignite, they can't be started up again for a second time on the launchpad. *Endeavour* would have to be rolled back to its hangar and outfitted with a new set of motors before another attempt could be made.[7]

This would be Arthur C. Clarke's last chance to see a blastoff from Florida up close. "Of course I'm disappointed. I've never seen a Shuttle launch before," Clarke told reporters. "But, of course, when things go wrong, one doesn't want to take chances." He did take solace from hearing the engines on the Shuttle, if only for a few seconds. The media had other things to talk about. One subject was whether Clarke would receive the Nobel Prize for the fiftieth anniversary of his most famous idea. Back in 1944, he wrote a paper in which he first envisioned communication satellites. He dreamed about spacecraft stationed in geosynchronous orbit around Earth that would bounce radio signals anywhere on the planet. Clarke joked with the press that if Cambodian Khmer Rouge leader Pol Pot didn't get the Nobel that year, he might have a chance.[8]

But there were also regrets about the apparent lack of progress the world had made in terms of space travel. Many people credit Clarke with dreaming up the direction the world is currently taking in terms of launching humans off the planet. In *2001: A Space Odyssey*, the character Heywood Floyd flew to orbit aboard a space shuttle operated by Pan Am. His destination was a space station operated by the Hilton hotel company.[9] In a case of life imitating art, PayPal entrepreneur

Elon Musk later built and launched privately owned rockets. The first boosters by his Space-X company were intended to carry cargo to the International Space Station for NASA. Later versions were hoped to launch people into space.[10] Another businessman, Robert Bigelow, may have been inspired by Clarke's idea for the fictional orbiting complex run by Hilton. Bigelow Aerospace worked to create a space hotel where paying customers would go to experience life in space.[11] Clarke's vision was of private business making money in low Earth orbit while the federal government's civilian space effort concentrated on the Moon and beyond.

When Clarke visited the Kennedy Space Center in 1994, NASA was still puttering along with the Space Shuttle. "I'm just sorry we haven't done more," said Clarke. "I've seen so much, but I feel bad for the young people who grew up on *Star Trek*. I'm sure they feel we should have been on Alpha Centauri by now." NASA wasn't visiting distant stars. The agency is still struggling to return to the Moon, let alone send people to face the dangers of a Mars mission. They have to overcome things like Moon dust first.

Selling a Moon Base to Congress

In January 2004, President George W. Bush announced the creation of Project Constellation. The Space Shuttle would be retired in 2010, and NASA would establish what the president called a "foothold" on the Moon. To the space agency, that meant a lunar base, and proponents immediately set about supporting the idea in Congress. Paul Spudis went before the U.S. House Science and Technology Committee in April 2004 to send a message, which was "of course we should go to the Moon." It was a viewpoint he had held for years. As a planetary scientist, Spudis had championed a permanent outpost on the Moon as early as 1990, as part of a White House advisory panel. He was later named deputy leader of the Science Team for NASA's Clementine mission in 1994. The spacecraft mapped regions of the Moon, including the poles, which later gained favor as a possible site for a Moon base. When he testified before Congress in 2004, Spudis painted a picture of the Moon as an obvious choice for future exploration.[12] "Because the Moon has no atmosphere and is a quiet and stable body," he said, "it is a premier place to observe the universe." Spudis envisioned observatories on the Moon where astronomers would have a unique and unobstructed view of the heavens. A Moon base, he contended, would also be a test bed for technologies for future missions, including a manned mission to Mars. "By learning space survival skills close to home," Spudis continued, "we can create new opportunities for exploration, utilization, and wealth creation. Space will no longer be a hostile place that we tentatively visit for short periods. It becomes instead a permanent part of our world."

By December 2006, NASA had released a basic plan for a new lunar base. Once the proposed Orion spacecraft was launched, the next step would be to build and utilize the descendent of the buglike lunar module that carried astronauts Gene Cernan and Harrison Schmitt on Apollo 17. Instead of just two people for Moon sorties lasting no more than three days, NASA intended the initial trips of the new Altair lander to carry four crew members on visits lasting seven days. The vehicles would carry barrel-shaped modules, which would be left on the surface to form an initial outpost. After that, a permanent Moon base would be built.[13]

Possible sites for the habitat included Shackleton Crater at the lunar South Pole. NASA planners favored a polar Moon base because the region was almost constantly bathed in sunlight. That meant a moderate temperature compared to the equatorial landing spots of Apollo. There was also the possibility of water in the form of ice in nearby craters, which the astronauts could harvest.[14]

Even this ambitious plan was comprised of baby steps compared to the challenges and dangers that humans will endure on Mars. It also had detractors who normally championed the mission of NASA.

The Not-So-Friendly Pilot of *Friendship 7*

The same U.S. House Science and Technology Committee that heard from Paul Spudis in 2004 gathered four years later to trumpet the accomplishments of NASA on its fiftieth anniversary. The list of those giving testimony in July 2008 included retired U.S. senator John Glenn, one of the original "Mercury Seven" astronauts who took America's first steps into space. Glenn later joined the crew of Space Shuttle *Discovery* for a research mission in orbit in 1998.

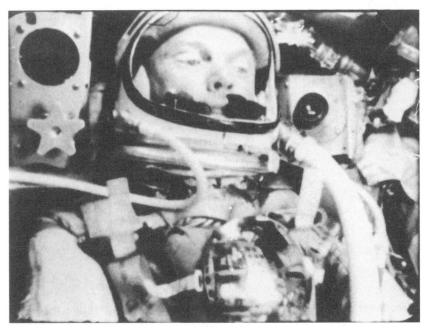

Astronaut John Glenn orbits Earth as the first American to circle the globe. Glenn would later join the crew of Space Shuttle *Discovery* on STS-95 as the oldest person to participate in a Shuttle mission. Courtesy of NASA.

The next Shuttle to launch after that carried up the first American-built section of the International Space Station. Instead of focusing on "sea stories" from his trip as the first astronaut in Earth orbit aboard the *Friendship* 7 space capsule, Glenn took direct aim at the plan for NASA's Moon base as a launching off point for a trip to Mars.

He didn't like it.[15]

"If that's what we're thinking about doing," Glenn told members of the committee, "it's going to be enormously expensive." The former lawmaker from Ohio criticized the Bush administration's plan to retire the Space Shuttle program, which, he contended, would leave America at the mercy of Russia. Until the new Orion capsule was ready to fly, U.S. astronauts would be dependent on rides from Russian Soyuz ships to visit the International Space Station. Glenn thought emphasis should be placed on the orbiting outpost. "The potential we have up there to learn new things is tremendous at a time when we're coming under additional global competition," he said.

Despite the opposition from NASA's most famous astronaut, the agency continued to press the need for a Moon base to pave the way to Mars. NASA administrator Michael Griffin chose the International Astronautics Conference in Glasgow, Scotland, on September 30, 2008, to make his case. He told delegates that Project Apollo had only scratched the surface of exploring the lunar surface. "The total human experience on the Moon is less than twenty-seven working days, on a world the size of Africa," he said.[16] As for a lunar base, Griffin claimed it would provide an endurance test for astronauts being trained for NASA's perilous first mission to Mars. The crew would be asked to simulate the rigors of a flight to the red planet by flying no farther than the Moon. "The experience would consist of putting astronauts on the space station for say six to seven months," Griffin told his audience in Scotland. "Then we'd take them to the Moon and ask them to survive there for nine more months with no further assistance than what they brought." The astronauts would then wrap up their trip by returning to the space station for a final leg lasting another half a year before returning to Earth. At no point would they receive any fresh supplies. Griffin contended that if humans couldn't survive that kind of mission close to the safety of Earth, then the outlook for NASA's first expedition to Mars would be grim. "The first crew to Mars would not come back," he believed.

Even the simulated flight Griffin envisioned wouldn't be possible without new technologies to create the proposed permanent Moon base. Many basic ideas were considered for the main habitat. Among them was a design for a crew cabin that would inflate like a balloon and another that would walk.

Have Base, Will Travel

NASA's new Moon base probably won't resemble the sprawling facility depicted in Arthur C. Clarke's *2001: A Space Odyssey*. If a group of designers succeeds, astronauts will live and work in something more closely resembling the imperial walkers that tromped around in the George Lucas film *Star Wars: The Empire Strikes Back*.

The agency's idea is a base equipped with legs.

NASA engineer Marc Cohen had a twenty-year career with the agency. In the 1980s, he worked as a designer at the Ames Research

Center near California's Silicon Valley to help design what would become the International Space Station. In the 1990s, he turned his attention to proposed Mars missions, including a robotic spacecraft that would gather up Martian soil samples and shoot them back to Earth on a return rocket. That mission never flew, and in 2002 Cohen found himself looking for a new project.[17]

He attended the World Space Congress in Houston, Texas, in October of that year. That's when a few beers at his hotel changed his life. He was approached at the bar by John Mankins, NASA's new chief of technology development. "He bought us a couple of rounds," says Cohen, "and he asked what I was up to. I said I had no assigned duties at that point. That's when he asked if I would be interested in working on something called HABOT."

NASA is cluttered with acronyms. HABOT was short for habitation robot, or a pressurized crew cabin equipped with legs for walking around on the Moon or Mars.[18]

Cohen had already written a paper about building big lunar rover vehicles that astronauts might use to drive to interesting sites on the Moon. Each would have a small, airtight cabin, so the crew members could work in a shirtsleeve environment without heavy spacesuits. It was the rescue scenario that kept complicating the idea. What would happen if the first rover broke down far from the permanent Moon base? Obviously, you'd send a second rover to retrieve the crew. Then, Cohen asked himself, what if the rescue rover got stuck? "If you follow the chain of logic," says Cohen, "then you're spending more and more of your resources in the rescue. So why not make the whole base mobile?" Suddenly, the notion of a Moon base that could crawl on its own was becoming more and more attractive.

The basic concept for HABOT was to send a series of six-legged compartments to the surface of the Moon. Astronaut crews would later touch down in lunar landers and climb inside the HABOTs. The result would be a caravan of mobile vehicles, each just slightly larger than the lunar modules of Apollo. One key point is that the HABOTs wouldn't be much heavier than their Apollo-based ancestors. Weight is crucial in the space program since it currently costs about $3,000 to launch a pound of hardware or people into orbit. That doesn't count the extra fuel needed for the trip to the Moon. The concept wasn't unlike

This is an artist's conception of the so-called HABOT mobile habitats first envisioned by the Ames Research Center. Designed to operate like the wagon trains of the American West, the multilegged crew modules would link up at day's end to form a base for the astronauts on the Moon or Mars. Courtesy of NASA.

the wagon trains that opened up the American West in the nineteenth century.

One added feature NASA was thinking about would be docking hatches at the front and back. At the end of the day, the HABOTs were supposed to line up and dock to each other to form one big structure where crew members could walk from compartment to compartment. "Some people liked the idea, but some were pretty skeptical," says Cohen. The idea of linking up multiple HABOTs allowed Cohen to include something in his design that came from his experience working on the International Space Station. Namely, he wanted a wardroom with a dining table for socializing after a day exploring the Moon. "It was an old navy concept, where officers would have a place to meet and eat," says Cohen. "Skylab back in the 1970s had a table and three chairs. A wardroom was included in what would have been the habitation module on the International Space Station. Even the Russian Salyut and *Mir* stations had a wardroom."

Creating a mobile lunar base would also help NASA solve a political issue that may complicate its efforts to return to the Moon, and maybe send crews to Mars—that is, to create a program that performs valuable science along with the initial achievement of setting foot on the lunar surface again. "I think that's the make or break argument for a lunar base," says Cohen. "Congress isn't going to fund this if NASA doesn't make a strong case for science." A mobile Moon base, its supporters contended, could see more territory.

Ironically, Cohen blames politics for causing the HABOT program to lose steam. There have been a lot of intriguing concepts at NASA, and not all of them come to fruition. During Cohen's career at the agency, he worked on the International Space Station, which is currently orbiting Earth. But he also spent time and effort on the Mars soil sample return vehicle, which was never launched.

When the Space Shuttle *Columbia* broke apart and burned up in February 2003, the White House changed NASA's mission. Cohen thought, at the time, that HABOT would be a likely casualty of the new mind-set. He contends that successful space projects involve a number of NASA's space centers around the country, and HABOT didn't do that. "The way things work at NASA," says Cohen, "is that a program has to have one or two big rockets which involves Marshall [the space center in Alabama], then the manned portion goes to Johnson [the space center in Houston], then you get industry involved so they stay in business." Cohen left NASA in 2005, leaving the idea of a six-legged mobile Moon base to other people. HABOT didn't die, but rather was reworked.

Reinventing the Wheel

Brian Wilcox had been thinking about walking Moon bases since 1998. His title at NASA's Jet Propulsion Laboratory (JPL) in Pasadena, California, was principal investigator into the concept. In other words, he had to make it work by creating and testing a set of spiderlike legs that would enable the compartment to crawl. Wilcox had heard the detractors who thought the idea was funny looking, but he believed in it. He pointed to history as an example. "Look at the ships of Columbus," Wilcox says. "The *Niña*, the *Pinta*, and the *Santa María* were all mobile habitats. And all of the missions that followed did the same thing. I think it's perfectly reasonable that the Moon and Mars will be explored by mobile habitats." Wilcox had seen the basic design concepts of HABOT coming from the Ames Research Center and had some definite opinions about it. "Nobody's going to build that," he believed.

JPL's answer was to add wheels to the legs. Instead of crawling like a spider, the revised design would mean astronauts would spend most of their time rolling from place to place, with the act of walking used for emergencies. HABOT's spindly leg design looked like the imperial walkers in the motion picture *Star Wars*. They may be nifty looking, but from an engineering perspective, mechanical legs were considered inefficient.

Wilcox and his colleagues also had compelling evidence that legs with wheels could be a lifesaver on the Moon or Mars. They had seen a hard lesson learned by the operators of the robot Mars rover named *Opportunity*.

The $400 million golf-cart-size explorer touched down on the region of Mars called Meridiani Planum in 2004. Scientists theorized the landing spot was the bed of an ancient ocean, which had long since dried up. The rover was equipped with cameras, sensors, and even geology tools. Its job was to photograph, scan, and scratch at the rocks it encountered. The robot was also perched on six wheels, which would propel it from place to place. At least that was the idea, until *Opportunity* ran into trouble in 2005. The rover was rolling toward Victoria Crater, a nearly 3,000-foot-wide hole punched in the surface of Mars.[19]

Opportunity's path was marked by ripples of sand, including a one-foot-high patch, which ground controllers aptly nicknamed "purgatory dune." The rover promptly sank in the sand to the tops of its wheels and got stuck. It took NASA five weeks to work *Opportunity* out of that jam, which wasn't lost on Brian Wilcox and the designers of the mobile lunar habitat. Their worst nightmare was seeing astronauts mired in a similar situation.[20] The concern drove the idea of putting wheels on each leg of the proposed mobile lunar base. "We wouldn't care if we got stuck in sand," says Wilcox. "We would have just locked the wheels, used them as 'feet,' and just tip-toed out."

The wheel-on-leg design was redubbed from HABOT to ATHLETE, which stood for All-Terrain Hex-Legged Extra Terrestrial Explorer. It

This is the base of the so-called ATHLETE mobile habitat envisioned by the Jet Propulsion Laboratory. The "legs" and "feet" of the base structure were intended to support a pressurized crew module on the surface of the Moon or Mars. Courtesy of NASA.

was yet another acronym for NASA's lexicon. Like scientists who flew *Mariner 4* to the red planet in 1965 or designed *Mars Pathfinder* in the 1990s, Wilcox found himself working to sell the idea of a mobile habitat to NASA.

Jeeps and Winnebagoes

Wilcox and proponents of the ATHLETE mobile base design sat in with NASA planners working on revisions to the agency's plan to return astronauts to the Moon. The main selling point to include the rolling lunar base was science. Initial artist conceptions for the complex had the astronauts living in one spot in a collection of pressurized modules. Critics say that would be fine, for about a month. NASA estimates that astronauts in roving vehicles could venture only about six miles from their base before having to return.[21] "Exploring ten kilometers of the Moon could take about thirty days," says Wilcox. "The scientific community wouldn't want to spend more than thirty days in any one location." Safety was another consideration.

When astronauts started landing on the Moon during Apollo, engineers learned about how the engines on the lunar module impacted the surface. "As you land, it tends to spray fine particles," says Wilcox. "The exhaust plume picks up these rocks and fine particles and propels them at a high velocity, like a bullet." Nearby crew modules could be hit and damaged by flying debris. A walking base could be relocated behind a ridge or some other form of protection far from an approaching lunar landing vehicle. Once the astronauts were on the Moon, the base could then roll toward the crew.[22]

That's when supporters of the mobile base concept coined the term "Jeeps and Winnebagoes" to describe the benefits of being able to move from place to place to more fully explore the Moon. "If you see a Winnebago towing a jeep down the highway, you know what's going on," says Wilcox. "They're going to park the Winnebago nice and level so they can stay for several days." According to Wilcox's scenario, the jeep would then be used to explore in a radial pattern from the base camp. "When they get tired of that area, they just move on to another one," he contended. The mobile ATHLETE base would be the Winnebago. The jeep would be a separate vehicle Wilcox's team calls the pressurized rover.

NASA's Proposed "Two-Seater"

Designer Marc Cohen began his association with the HABOT mobile base concept after writing a paper on a updated Moon rover, similar to the lunar buggies used by the last three crews of Project Apollo. When that project evolved into Brian Wilcox's ATHLETE design, the rover came along. The idea was to build a vehicle with a small, airtight compartment where crew members could roll around on the Moon without having to wear heavy spacesuits. The Apollo astronauts wore their suits when using the older lunar rovers in the early 1970s.[23]

NASA added wording to its lunar plan saying that mobile compartments like ATHLETE should be considered. That led to tests of prototypes at Moses Lake, Washington, in the spring of 2008. Two modules, each equipped with six legs with wheels, rolled over the dunes and hills around the lake to simulate the kinds of terrain a mobile base might face on the surface of the Moon.[24]

Brian Wilcox admits a mixture of ideas is the likely outcome for the final design of NASA's proposed Moon base. By that, his team thinks the space agency will choose to establish a permanent Moon base with the capability to explore either with rovers or mobile compartments. If that turns out to be true, NASA might turn to another innovation. It's a base that inflates.

Life in a Balloon

In 1952, an appendix was issued for a novel written by rocket pioneer Wernher von Braun. It spelled out how to build the inflatable crew modules the book envisioned as a way to send people to Mars. Science fiction took a step toward reality in 1960 with the launch of the first in a series of NASA's ECHO satellites. The spacecraft contained balloons, which were filled with air after reaching orbit. Their shiny Mylar surfaces were used to bounce radio and television signals from one spot on the globe to another.[25]

Not to be outdone, the Soviet Union utilized an inflatable airlock on its *Voskhod 2* spacecraft, which launched on March 18, 1965. Cosmonaut Alexei Leonov used the chamber to exit his capsule and perform the world's first spacewalk. The stunt almost ended tragically when Leonov found that his spacesuit, filled with air, was too stiff to allow him to close the hatch once he reentered the airlock. He had no alternative but to open a valve on his spacesuit to bleed off enough air to allow him to get back inside.[26]

Despite the near disaster, the notion of using inflatable spacecraft gained more and more popularity.

When Space Shuttle *Endeavour* blasted off on STS-77 in 1996, one of its payloads was the SPARTAN-201 satellite. The small, boxy spacecraft had a silver inflatable antenna the size of a tennis court.[27]

In 2003, NASA launched the two Mars rovers, *Spirit* and *Opportunity*. Each could bounce to a halt on the surface of the red planet with inflatable airbags that surrounded the robots and cushioned their touchdown. That airbag system was built by a NASA contractor in Delaware called ILC Dover.[28]

This is the inflatable prototype of a NASA Moon or Mars base habitat module, designed and built by ILC Dover. It was tested at McMurdo Station at the South Pole. Photo courtesy of ILC Dover.

Despite the use of inflatable technology in space, Dave Cadogan spends a lot of his time trying to convince people that a base on the Moon that blows up like a balloon isn't a crazy idea. His job is director of research and technology at ILC Dover. The International Space Station came close to having an inflatable module as its crew compartment. The project was called Transhab, and featured a 30,000-pound module with 12,000 square feet of space. The program was discontinued in 2000, but it did lay the groundwork for possibly using inflatable modules to build a base on the Moon or Mars.

NASA's vision for the lunar complex means using metal compartments built on Earth and carried up on rockets. Proponents of inflatable modules say their technology makes sense for the next step, which is building a permanent base.[29] "A number of 'tin cans' will get the job done in the short term," says Cadogan. "But when you need more volume to keep the crew sane over longer periods, that's when you want to look at inflatables."

Cadogan admits getting people comfortable with the concept is a tough and ongoing job. "There are perception problems with inflatable structures that people have," he says. "And it's amazing how pervasive it is." Supporters point to common items like automobile airbag safety systems and the inflatable escape slides on commercial aircraft. Using this idea in space, they contend, is just a logical progression of current technology.

The compartments ILC Dover has in mind would be thirty-two feet long, with windows and an airlock so astronauts could move in and out to perform work on the lunar surface. Early concepts included the development of the outer skin, which would need to be resistant to the constant bombardment of ultraviolet light from the Sun and the extremes of heat and cold on the airless lunar surface.[30]

NASA and the National Science Foundation liked the idea enough to propose a prototype that could be tested under conditions as close as possible to the planet Mars. That meant building the inflatable room and shipping it to Antarctica.[31]

To the Bottom of the World

McMurdo Station looks like a cross between a college campus and an Alaskan trading post. It's located on the ice shelf of Ross Island in Antarctica, the southernmost spot on the continent that is still accessible by ships. Established in 1956, it has grown from a sparse outpost to a small town with an airport, a helicopter pad, and about a hundred buildings complete with water, sewage, electricity, and telephone lines.[32]

Jason Bryenton worked at McMurdo as a research associate. His job was to monitor scientific equipment for various organizations doing studies in the Antarctic and to ensure these groups got the data they wanted. He liked it so much he applied for a second one-year assignment at McMurdo and got it. "We're surrounded by mountains down here," he says. "There's an active volcano that almost always had a plume coming from it. We get to see a number of different types of wildlife like seals, penguins, and orcas. It's a truly amazing place." McMurdo's population can go from around 200 people during the height of winter to more than 1,000 during the summer when the base is more accessible.

Being at the bottom of the world, the change of seasons can appear odd. During winter, it's almost constantly dark. Summertime is the opposite, with the sun never really rising or setting. It appears to move in a circle overhead. "I like to joke that it's two o'clock in the afternoon all day long," says Bryenton. "Some people find it disruptive. But I like it. It adds to the experience."

From ILC Dover's perspective, McMurdo was also the best spot on Earth to test its inflatable Moon and Mars habitat. The severe cold gets as low as minus 112 degrees Fahrenheit, and the winds are brutal. Also, when the ozone hole opens up, the stream of ultraviolet radiation from the Sun is the most intense on Earth. With no atmosphere on the Moon and a thin air layer on Mars, radiation is a major concern for astronaut crews. The questions about the inflatable habitat also went deeper than whether it could survive under the harsh conditions near the South Pole. NASA wanted to know how easily the inflatable room could be transported from place to place, filled with air, and maintained.[33] "We're also testing the team's response to getting it there," says Cadogan. "How do they fix it when something goes wrong? What they can do and cannot do inside the structure, like dropping things."

Bryenton was assigned to be the compartment's "keeper" at McMurdo. He was there when it first arrived and was set up. The finished product was a green cylinder covered with white spots. It was twenty-five feet long and eighteen feet high. "I was expecting something that was configured to go on the Moon or Mars," he says. "There was no airlock to keep it safe from a reduced atmosphere environment. It just looked like a big tent." The inflatable compartment was equipped with cameras and sensors to monitor its response to the Antarctic environment. "I've been inside it a number of times," says Bryenton. "There's a table with some of the instruments on it, an air pump to keep it inflated, and a couple of translucent windows to let some light in." The process of watching over the test habitat included on-site checks twice a week, regular leak inspections, participating in occasional teleconferences with engineers at ILC Dover, and ensuring there are no unauthorized visitors. "We're not talking about penguins or anything," says Bryenton. "People are just curious." Engineers had hoped that the tests would expose the prototype to a number of conditions all at once, which had not happened with previous NASA hardware. The concern was over how the different elements of cold, wind, and ultraviolet light exposure might cause the material in the habitat's skin to break down. ILC Dover points to the protective covering for the Hubble Space Telescope as an example. "NASA tested it in a vacuum chamber," says Cadogan. "They tested it against UV radiation and other forms of radiation, but not all at once." When astronauts first visited Hubble to make repairs in 1993, the thermal covers were found to be tattered.[34]

Inflatable crew quarters for the Moon or Mars might seem like a novel idea, but how far does the technology need to progress before it's durable enough to withstand the rigors of a space mission? Cadogan says not much, assuming NASA is prepared to invest the time and money to finish development. The list of possible improvements includes a skin that contains sensors to warn astronauts if a leak has occurred. "The crew can spend a lot of time looking for leaks," says Cadogan. "We want smart structures that can tell the astronauts where a leak is, so they can slap a patch on it." Another innovation might be an outer covering with the ability to seal small punctures on its own without human intervention.

The issue for NASA comes down to time, money, and the decision to proceed with any particular technology, whether it's an inflatable base, one that walks, or the most traditional "tin can" form of crew module.[35] One of the decision makers is Geoffrey Yoder, director of the Integration Office at NASA's Headquarters Exploration Office. He sees advantages and drawbacks with both the inflatable and mobile base designs. The idea of a crew cabin equipped with wheels or legs is good, unless there's a compelling reason why the astronauts should stay in one place.[36] "If you talking about generating oxygen from lunar soil," says Yoder, "then you want to keep it stable, so you're not picking it up and moving it around." The concept of inflatable crew cabins is intriguing as well, but Yoder thinks that building whole bases that way might not be practical. "It may take a combination of hard-shell compartments and inflatable modules to do it," Yoder believes. "The Antarctica test program is to see what it can do, but the whole thing probably wouldn't be inflatable."

The debate over how to achieve space goals like this goes back to the days of Apollo. Astronauts blasted off in Saturn rockets and orbited Earth before launching on a path to circle the Moon. Two crewmen then climbed into the enclosed and pressurized cabin of their lunar landing module for the trip down. During the early planning process in the 1960s, there were proponents of an alternate path. The other plan was to launch a rocket directly from Earth to the Moon without orbiting either body.[37] Instead of a two-man lunar module, some engineers favored using a small "Moon hopper," which was like a rocket-powered miniplatform. It would carry one man in a spacesuit to the surface for solo exploration. After much argument, NASA settled on the now-familiar Apollo hardware and strategy to put astronauts on the surface. The next-generation lunar base and Mars missions will probably follow that same process.

The agency has years to go before a long-term Moon base might be needed. Along with hardware, NASA will need people to make the trips and perform the expeditions on the Moon to test the tools needed for the more perilous missions to Mars. As the U.S. space program has evolved, so, too, have the qualifications for the men and women who are chosen to be astronauts. As NASA changes, the so-called right stuff is changing as well.

the new
"right stuff"

What kinds of people should NASA send as astronauts to Mars? That's a question that's haunted the agency since the mid-1990s. Pioneers faced the western frontier on a volunteer basis. NASA will pick crews to go by a process that even current astronauts consider mysterious and arcane. Ironically, the subject of who might be picked to go to Mars was a forbidden topic within the space agency. Members of the elite astronaut corps could get into trouble for just talking about it in public.[1] The U.S. space agency's job was to fly missions with its fleet of Space Shuttles and to work on the Cold War version of its space station called *Freedom*. Plans for the orbiting complex would later evolve into the International Space Station. That was it. "Mars was pretty much verboten," recalled astronaut Scott Horowitz, who considered himself among the heretics who thought a lot about the idea of a Mars mission. He joined the astronaut corps in 1992 and throughout his career busied himself with flights to repair the Hubble Space Telescope and two trips to the International Space Station. Still, Mars was lurking in the background.

That's not to say that NASA headquarters wasn't doing some preliminary planning for a possible Mars flight during that time. This underground movement involved some well-known names within the agency. "The early architecture for a Mars flight was done by a fellow named Doug Cooke," says Horowitz. "And the exploration bureau was headed by Mike Griffin." Cooke later became NASA's deputy associate administrator of exploration, and Griffin was appointed by President George W. Bush as the agency's chief administrator in 2005. "It's fun to go back in history to see who was doing what," says Horowitz.

Astronaut Scott "Doc" Horowitz undergoes a suit check prior to boarding Space Shuttle *Columbia* on mission STS-75. The crew would deploy the Italian tethered satellite on a long cord, in part to generate electricity by friction of the tether with molecular oxygen. Courtesy of NASA.

One detail Cooke was interested in was the makeup of the ideal astronaut crew for a proposed mission to Mars. Up to that point, the people waiting alongside Horowitz for a Shuttle mission included military fighter pilots, medical doctors, electrical and mechanical engineers, and astronomers, among others. Mars would have different needs.[2] NASA's days of short space missions, like vacations, were coming to a close.

The qualities that set astronauts apart from the rest of humanity, the so-called right stuff, have evolved over the years as NASA's mission matured. Project Mercury sent up one-man capsules to prove that humans could live and work in the strange environment of space. That called for candidates with strong backgrounds as test pilots, like John Glenn, Gordon Cooper, and Wally Schirra. As the United States planned to send people to the Moon during Project Apollo, astronauts with engineering expertise were selected. The first scientists were chosen to fly in 1966. By the time the Space Shuttle was launched in 1981, most astronauts were systems operators with jobs like working the spacecraft's lanky robotic arm or performing spacewalks.

The trend of current skills becoming outdated is a familiar one at NASA. Work aboard the International Space Station, for example, made previous achievements obsolete. Spacewalkers outside the orbiting outpost routinely did complicated tasks like changing the bearings on sets of solar electricity panels. The job enabled the long, winglike arrays to turn to track the Sun and generate power. Intricate work like this was unheard of among the astronauts even just a few years earlier.[3]

In 1997, Scott Horowitz was assigned to be the pilot aboard Space Shuttle mission STS-82. During that flight, spacewalkers ventured outside in thick white spacesuits to install ten new instruments on the Hubble Space Telescope. Compared to the work going on aboard the International Space Station a decade later, his Hubble trip looked quaint.[4] "If you had laid out that kind of EVA timeline [on the station] just five years before," says Horowitz, "you'd have been laughed out of the room." Now it was just another day at work. Going to Mars, obviously, will require NASA to stretch even further.

With Mars as the objective, NASA was readying itself for another fork in the road. Doug Cooke asked Scott Horowitz to help narrow the focus of who should be a Mars astronaut candidate. It was clear that traveling the thirty or so million miles would require people and technologies the agency didn't have.[5]

Cooke and Horowitz put their heads together to try to define NASA's future needs. Their solution was a survey. It was tacked to a bulletin board for current Space Shuttle crew members to answer. It listed the various types of people who might be included on a mission to Mars. The astronauts were asked to rank the thirty or so jobs from the most important to the least. With almost no exceptions, the respondents agreed on the most vital crew member.

They picked the mechanic.[6]

The result made perfect sense to Scott Horowitz. "Everybody knew there was no way you're doing a mission to Mars without the ship breaking down," he says. "The mechanic was someone your life depended on." That could be someone like astronaut Don Pettit, for example.

The "Mr. Fix-it" of the International Space Station

Don Pettit first flew aboard Space Shuttle *Endeavour* in December 2002 to begin his three-month-long stay aboard the orbiting complex. He was one of the few astronauts to get a laugh during the traditional press conference during launchpad practice one month before blastoff. He and his crewmates gathered in a small auditorium at the Kennedy Space Center press site to take questions. A reporter noted that his mission would mean being in space over the Christmas holiday. Pettit was asked if he had any plans to get presents for his family. "I guess I could get something at the Kennedy Space Center gift shop," he quipped.[7]

Pettit and his station crewmates, astronaut Ken Bowersox and Russian cosmonaut Nikolai Budarin, remained aboard the laboratory when *Endeavour* pulled away for the trip back to Earth. The station was only two years old during their stay and only a fraction of its completed size. Crew modules, including the U.S. laboratory and the Russian-built crew compartment, were topped by a single set of the gold winglike solar panels on a tall mast.[8]

Along with routine maintenance on the outpost, Don Pettit became renowned for his ability to repair broken equipment with what became known as his "strange tool bag." "He was up there soldering resisters and fixing things that didn't work," says Horowitz. "I mean there's no way you could ship new stuff up from the ground otherwise."

Pettit and his crew were also aboard the station when Space Shuttle *Columbia* disintegrated and burned up during reentry in 2003, killing the seven astronauts on board. The Shuttle fleet was grounded while NASA investigated the accident. That meant the space station crew had only one way to return to Earth at the end of its mission in May 2003. The three men would have to come down in a Russian Soyuz space capsule. In fact, they would be the first people to land in the latest model, called the Soyuz TMA-1. During reentry, the capsule's autopilot failed to take a smooth path down through the atmosphere, where the crew would experience three times the force of gravity in the cramped vehicle. Instead, the capsule plummeted on a much steeper course, generating nine times normal gravity on the astronauts' bodies. To make matters worse, the crew landed 300 miles away from their landing tar-

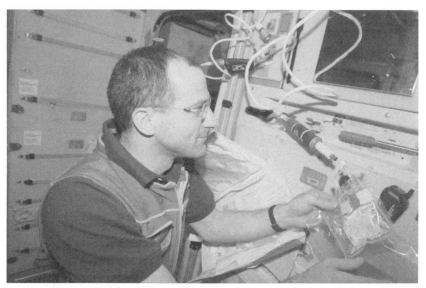

Astronaut Don Pettit served aboard the orbiting outpost following the loss of Space Shuttle *Columbia* in 2003 and later flew on the Shuttle mission to work out the kinks on the Space Station's water recycling system that makes urine drinkable. Courtesy of NASA.

get. After that, Soyuz vehicles were equipped with satellite telephones so crew members in a similar situation could simply "call" for help.[9]

When Pettit returned to the outpost in 2008, he was asked how it felt going up and back on Space Shuttle *Endeavour* compared to his Soyuz return five years earlier. "I'd compare it going economy class to going business class," he says. Pettit's tinkering skills would also be put to the test with an experimental system to purify water for the station. The hardware drew a lot of media attention because it would take sweat and urine from the crew and clean it for later consumption. It was called the Water Recovery System, and its success was crucial if NASA was going to increase the number of people living and working full time on the complex. Designers promised the unit could generate up to six gallons a day for the crew. The alternative to recycling water was to ship it up on rockets, which was complicated and costly.[10]

The location of the new unit on the space station left little doubt as to what was going on. The complex's new toilet was positioned on one side of the water purifier, and the galley was on the other side.

Wastewater from one location would be cleaned and funneled to the kitchen for use by the crew. The commander of the outpost during the installation of the system, astronaut Mike Fincke, defended the idea of drinking reconstituted wastewater. "Actually, it goes through such a process that it's no longer urine," he says. "It's just clean water, probably cleaner than most people's tap water. I'm not afraid to drink it." The press remained leery, and that part of the mission generated a lot of jokes. Prior to liftoff at the Kennedy Space Center, a member of the public affairs staff wrapped up a televised interview by taking a sip of purified water from the system. The act was met with grimaces from the assembled reporters, but no ill effects were seen.[11]

Endeavour's launch drew more than the usual number of onlookers along the Indian and Banana rivers east of Orlando, Florida. Those waterways separated mainland Florida from the strip of land along the Atlantic, which was home to NASA's Kennedy Space Center. The media billed the blastoff as the final night launch before the end of the Space Shuttle program. Daytime liftoffs include the rumbling of the Shuttle's solid rocket boosters as the vehicle climbs to orbit on a column of fire. Similar events at night were much more dramatic. Pettit and his six crewmates donned matching orange flight suits, complete with survival gear and heavy helmets, for the trip to the launchpad.[12]

The crew of STS-126, which delivered the new water system, included the usual lineup of astronauts. Commander Chris Ferguson would fly the Shuttle, assisted by pilot Eric Boe. Spacewalkers Heidi Piper, Shane Kimbrough, and Steve Bowen would work outside the station. Astronaut Sandy Magnus operated the robot arm. Don Pettit's primary job was to get the water purifier working. NASA knew the system would likely not cooperate at first. So the agency needed a "master tinkerer" to get the job done.[13]

Endeavour's takeoff at just before 8:00 p.m. was as brilliant as an instantaneous sunrise. The vehicle parted overhead clouds, causing ripples in the sky like a stone tossed into a pond. Minutes later the spacecraft was reduced to the appearance of a slowly moving star that seemed to be dropping down into the ocean. In reality, *Endeavour* was just going into its first orbit around Earth.

The water-reclamation hardware was housed in two black space station equipment racks, each about the size of a telephone booth. Silver distillation drums and filters, connected by a maze of metal pipes, were intended to remove debris like lint and hair, as well as bacteria, salts, and other biological contamination from human wastewater. Designers promised the system would purify 95 percent of the urine and sweat from the crew members and provide about six gallons of fresh water daily. Maintenance on the racks would be limited to changing filters once they were clogged with impurities. Each of the racks containing the new gear was latched inside a barrel-shaped Italian cargo module dubbed *Leonardo* for the trip to orbit. The warehouse module was locked inside *Endeavour*'s cavernous payload bay. Once the space plane docked to the space station, the astronauts used the Shuttle's robotic arm to lift the portable compartment out and snap it into the outpost like a temporary crew module. At that point, the purifying gear and all the other supplies inside *Leonardo* were moved into the station.[14] Five days later, the unit was switched on, and instead of a steady stream of water, the astronauts received a steady stream of "error" messages. The equipment repeatedly conked out.

NASA managers hedged their bets that this would happen. The filtering system's first six months aboard the station were meant to iron out the bugs. A Space Shuttle mission was planned during that time to bring up any extra parts needed to get everything running properly. Still, the agency's top brass couldn't hide their desire to the get the system working. All that was needed was for the process to work for just four hours. That was long enough to get a water sample for return to Earth. Engineers wanted to double check that the filters and the pipes connecting them to the toilet and galley didn't allow pollutants to leach into the water supply.[15] There were political considerations as well.

Many nations had contributed to the building of the International Space Station, including Japan, Canada, and the members of the European Space Agency. Only American astronauts and Russian cosmonauts had made long-term trips aboard the outpost following the start of construction in 1998. The only exception was when French general Leopold Eyharts spent two months in orbit after the installation of Europe's *Columbus* laboratory on the station. It was still far less than his colleagues at NASA and the Russian space agency. Going from the standard crew of three up to a total of six would provide opportunities for foreign partners to fly. It would also allow the people on the station to do more science along with maintaining the complex. Sending bigger crews hinged on the smooth operation of the Water Recovery System, which had to be tested before the crew could use it.[16]

The Tinkering Begins

The most high-profile work during Pettit's Shuttle mission were the spacewalks outside the station. Heidi Piper, Shane Kimbrough, and Steve Bowen would work to loosen a stuck set of solar panels with grease guns and cloth wipes coated with lubricant. Instead of turning to track the sun, the panel's swiveling joint was grinding metal on metal. The work was punctuated with the televised loss of a tool bag containing two of the grease guns. Piper admits she failed to secure the bag with tethers, and it floated off out of reach. "Great!" Piper was heard to say in frustration over the communication link. It was a high-profile mistake, but similar ones could be traced all the way back to the very first spacewalk in NASA history. Astronaut Ed White ventured outside the Gemini 4 spacecraft on June 4, 1965. In the process, a silver spacesuit glove was seen drifting away from the capsule.[17]

Inside the space station, Don Pettit continued to try to coax the water purifier back to life. Between work sessions, he defended the concept during televised press interviews beamed to the ground. "You have to remember, this equipment is serial number 0-0-1 of a brand new system that we're testing out on the station," he said of the balky system. "You have to expect a few hiccups." Mission control even extended the Shuttle flight by a day to give the astronauts more time to work.[18]

After days of trial and error, the fresh water finally flowed for the minimum four hours that NASA wanted. Pettit said the astronauts wouldn't be toasting the system's success with purified water just yet. "The engineers don't want us to drink any of it," the astronaut said. "We're bringing all the samples back with us. They want a few weeks on the system before giving their blessing for operation for human consumption, which is a smart thing to do." Between bouts with the water purifier, Pettit further impressed his colleagues by devising a way to have his beloved morning coffee without having to squeeze the fluid from a plastic bag into his mouth. He simply tore the plastic cover off a manual, folded it into the shape of a teardrop, and drank his coffee normally by using the surface tension of the fluid to keep his beverage from spilling out.[19]

Members of the astronaut corps were wowed. Pettit's versatility in orbit even prompted thoughts about changes in how NASA picked future astronauts.

Scott Horowitz started applying to the U.S. space agency in 1978 when NASA announced the very first class of Shuttle astronauts. He was a distinguished fighter pilot for the U.S. Air Force, having previously earned a doctorate in aerospace engineering from Georgia Institute of Technology. Horowitz's educational background earned him the nickname "Doc," which followed him throughout his career as an astronaut. He was turned down the first time around, but kept reapplying throughout his years in the military. "I think they just got tired of hearing from me and said 'yes,'" says Horowitz. He received a phone call from the Johnson Space Center at 2 a.m. while he was stationed in Germany. It was an invitation to come to Houston, Texas, later that week for a series of interviews to become a member of the astronaut corps.[20]

The prospect of sitting down in front of veteran astronauts and scientists at NASA put Horowitz in a near panic. He went through his old college notebook memorizing various equations from his aerospace courses in case the subject came up. He sat down with officials at the Johnson Space Center, and one of the first questions they asked was "What instrument did you play in the high school marching band?" The trombone, Horowitz responded. The exchange got him thinking about the criteria the agency was using to pick prospective Shuttle crew members.[21] Many of the people competing with him had impressive educations and wide experience. Then something became more and more obvious to him. "They were picking people for a camping trip," says Horowitz of the rigors that astronauts would face during the two-week mission in orbit. "If you could put seven of these people in a Winnebago, weld the doors shut, and ensure they're still talking to each other after seven days, you'd have a good crew." That may have worked fine during the early years of the Space Shuttle, but sending people on lengthy and dangerous missions to Mars kept harkening back to the versatility of astronauts like Don Pettit. "I'd put future astronaut candidates in a room with a V-8 automobile engine," says Horowitz, "and tell them to take it apart and put it back together." Others within the space agency were thinking along slightly different lines.

Geoffrey Yoder, head of the Exploration Mission Directorate at NASA headquarters, is also looking at a new type of astronaut as the era of the Space Shuttle draws to a close and flights to the Moon and Mars are being designed. His main point is hiring people who act with more autonomy. Crew members aboard the Shuttle and the International Space Station currently barrage mission control with messages to confirm their next action or seek guidance from Houston. Yoder's point is that future astronauts on the Moon or Mars need to act on their own. "By autonomy," he says, "I mean not prescribing everything down to the minute, but rather prescribing ground rules and then allowing astronauts to make educated decisions on their own." The main reason is the time delay in radio messages from Earth. It takes three seconds for a message from Earth to reach the Moon, and twenty minutes for a transmission to make the round trip to and from Mars. What Yoder is describing is something the Russian space program has done for years. Instead of choreographing space missions in painstaking detail, cosmonauts are trained in certain skills and then called upon to react to problems and new objectives as they come up. In that way, future astronauts may all be equipped with their own version of Don Pettit's "strange tool bag" to go off and solve things on their own.[22]

Pettit's knack for fixing equipment without losing his sense of humor means more than just getting the job done on a foreign planet. It was also cited as an example of the kind of psychological resourcefulness the agency will need during hazardous and frustrating missions to Mars.[23]

Lawrence Palinkas has spent much of his career studying what makes astronauts tick. He's a professor of anthropology and social work at the University of Southern California. He has also conducted studies on astronaut psychology for NASA beginning in 1986, which was marked by the *Challenger* accident. He began his work related to spaceflight by studying data taken from personnel sent on long missions to the Antarctic. The impact of long isolation at the bottom of the world, and working in close proximity to the same people for months at a time, is seen to be a close example to what astronauts will face on long flights to the Moon or Mars. In Palinkas's mind, the challenge is how NASA crew members inside a confined spacecraft will cope with their environment and each other.

A flight to Mars could mean six months on the trip from Earth, one or two years on the surface, and then six months on the return voyage. "There's a point where the last thing you want to see are the same faces you've seen over the last six months," says Palinkas. "Even the way they chew their food will get on your nerves." The situation researchers envision isn't like on television or the movies where an astronaut has a nervous breakdown and attacks his or her crewmates. However, the men and women on a Mars flight could face burnout or unexpected stress from Earth. "There could be a message from home that a loved one has died," says Palinkas. "On the one hand, your crewmates could be a source of support and strength, but also someone might want to be isolated from others to deal with their grief." Missions to the Moon or Mars will last much longer than anything NASA has dealt with before. Even trips aboard the International Space Station last only about six months. However, the agency did appear to recognize the problem of astronaut interaction during long flights in the early days of the Shuttle program. "NASA realized that working with someone you don't particularly like is different on a fourteen-day Shuttle mission, compared to working with them six months," says Palinkas. "We have to identify who has certain social needs or skills for long-duration trips compared to the shorter ones." The Russian space program recognized this dynamic and had a unique solution for cosmonauts in training for long flights on the older Salyut space stations or *Mir*. The crews would all be put in a car and told to go on a road trip together. "Going cross-country on those Siberian roads and so forth," says Palinkas, "is a good analogy to what they'd experience while working in space."

Psychological problems that appear halfway on a long flight to Mars are considered to be a big showstopper for NASA, even ranking as high as radiation exposure or the loss of bone and muscle mass due to the lack of gravity. One way to avoid this kind of breakdown is to look at the kind of person who is chosen to go.[24]

Superman or Clark Kent?

Lawrence Palinkas uses the analogy of the comic book superhero Superman and his mild-mannered alter ego, Clark Kent, to draw the line between what the stereotypical astronaut candidate was like in the past and what he or she may have to be in the future. "Superman tends to change his environment to suit his needs," says Palinkas, "while Clark Kent kind of rolls with the punches."

Astronaut Scott "Doc" Horowitz is a good example of the classic overachiever who was once in the employ of NASA. He was in the sixth grade when Neil Armstrong and Buzz Aldrin walked on the surface of the Moon during Apollo 11 in 1969. The event sent him into deep concern over his own future career as an astronaut. He wanted to go to Mars, but NASA appeared to be moving too fast. "Looking at where we were in 1969," says Horowitz, "I thought we'd be going from here to here to here in space and we'd be on Mars in another five or ten years." Horowitz was worried about missing his chance to walk on the red planet, but NASA's plans for a 1984 Mars mission quickly sputtered out due to budget cuts. Still, Horowitz's goal of joining the elite astronaut corps had been set into motion. He earned his doctorate in aerospace engineering in 1982. The next step was to join the military for experience flying fighter jets, which NASA seemed to like in astronaut candidates.[25]

There were no pilot slots open, so Horowitz went to work as a research scientist at Martin Marietta in Georgia. He had warned his boss that he might be on the job only a short time if the U.S. Air Force offered him a pilot's slot. Two months later the Pentagon made a phone call, and Horowitz was quickly on his way to flying F-15 fighters, and later T-38 training jets as a test pilot at Edwards Air Force Base in California. That base was where many of NASA's original Mercury astronauts had worked and where Chuck Yeager broke the sound barrier in 1947 aboard the Bell X-1 rocket plane, nicknamed the "Glamorous Glennis." Horowitz kept applying to NASA and was eventually accepted. Following his career as an astronaut, Horowitz opened his own aerospace company and helped NASA design the Ares-1 rocket that was intended to carry the next-generation Orion capsule to orbit.[26]

His dizzying résumé isn't unique among astronauts. And considering what's expected of Space Shuttle crew members, it's not surprising that this kind of person is the one who makes the grade. "These are the people who volunteer to be astronauts," says Palinkas. "They're also the ones who have the stamina necessary to withstand the training that we require of astronauts." But he questions whether that kind of temperament will work on a long-duration flight to Mars.

Stereotypical space flyers who have impressive military records, multiple university degrees, climb mountains, or hold patents on devices they designed and built all are accustomed to being successful. "That's how they got to be astronauts," says Palinkas, who thinks an easygoing mind-set might be desirable among the men and women selected to trips to Mars. "There are some features of the so-called right stuff astronauts that might not be as adaptive to a long-term space mission compared to the early days of the program," he asserts.

NASA might resolve the problem of selecting the right people for the rigors and dangers of Mars flights with computer programs that simulate frustrating equipment failures, which might gauge which candidates are best able to cope.[27]

The Loneliest Job at NASA

One thing that's likely not to change when people venture to Mars is the burden on the shoulders of the mission commander. NASA coined the term "commander" when the agency moved from the one-man Mercury capsule to the two-man Gemini spacecraft that helped make the Apollo Moon missions possible. On Gemini, a rookie astronaut would act as backup to the experienced crew member in charge of the flight. No one, even among the first-time fliers, wanted to be referred to as the "copilot." NASA remedied the problem by calling the lead astronaut the "commander" and the new person the "pilot."[28]

Being commander is described as the toughest job at NASA, and the loneliest. Take Andrew Allen, for example. The Marine Corps veteran was selected by NASA to become an astronaut in 1988, just two years after the *Challenger* disaster. Allen flew on three Shuttle missions; his last was as commander. Upon his retirement from the astronaut corps,

Astronaut Andrew Allen prepares for liftoff as commander of STS-75, which was the second mission to deploy the Italian tethered satellite. Courtesy of NASA.

Allen worked as a manager for United Space Alliance, a NASA contractor that prepared Space Shuttles for liftoff. When *Columbia* was being readied to carry the Chandra X-Ray Observatory to orbit in 1999, Allen was at the Kennedy Space Center press site in a gray business suit, which reflected the flecks of gray in his black hair. He remembered, distinctly, the difference between being a pilot on his first two missions and when he commanded his final flight.[29] His first trip, in 1992, carried up a 700-pound Italian satellite attached to a fifteen-mile-long cord to be reeled out like a bob on a fishing line from a large drum in *Atlantis*'s open payload bay.

Allen and his crewmates, Jeffrey Hoffman, Marsha Ivins, Claude Nicollier, Franklin Chang-Diaz, Franco Malerba, and Commander Loren Shriver, watched as the white spherical spacecraft drifted out into space, attached only by its thin cable. The whole thing made it out only 700 feet before the cord became snagged. The crisis created scenarios among the press of the Shuttle becoming tangled in the white cable.[30] While Allen was pilot aboard his mission, the tough decisions on what to do fell to Shriver, who was commander. "When you're the pilot on your first mission, or second or third for some," recalled Allen, "you know that if you get into a bind, the commander is going to be there to take charge and make everything work, and keep everybody safe, so you have that backup." Shriver's choice was to reel the satellite back in and head back to Earth.

Allen's second mission was also as pilot in 1994. On that voyage, the astronauts spent two weeks in orbit performing experiments aboard *Columbia*. Two years later NASA called upon him to serve as commander of STS-75. It was the second flight of the Italian tethered satellite that had failed during Allen's first flight in 1992. He would be joined by three crew members from the previous try, Hoffman, Chang-Diaz, and Nicollier. There would also be two Italian astronauts, Maurizio Cheli and Umberto Guidoni, and rookie pilot Scott Horowitz. The difference now was that it was Allen's responsibility to make the mission of *Columbia* work as planned.[31] "As commander, you 'sign' for the spacecraft," says Allen. "You sign your name that you're going to keep it safe, and that you're going to do everything humanly possible to make the mission a success, and, if anything comes up, you're going to take care of it and bring everybody home." If everything goes well, the commander stays in the background while the other astronauts go about their work, conducting experiments, doing spacewalks, and doing whatever else the flight requires.

Andy Allen's first trip as commander didn't go as planned. The Italian satellite was reeled out from *Columbia* routinely, until the cable snapped and the satellite went sailing off into space, lost forever. "So all of a sudden, you don't have a backup," said Allen of being in charge of the mission. "It brings a little more stress, but you work the mission and your crew, and make sure they work well together and they make

Space Shuttle *Columbia* blasts off on the second mission including the Italian tethered satellite. The flight didn't go as planned when the deployment of the satellite, being reeled out on a long white cord, ended in failure when the tether snapped. The white spherical satellite sailed off and was lost. Courtesy of NASA.

you look good." The tough part of the mission was over for most of the astronauts, but not Allen. The perilous reentry through the atmosphere was yet to come, along with the unpowered glide back to the Kennedy Space Center. One hand would be on the Shuttle's control stick through the one chance at a safe touchdown, and it was Allen's. "The entry is unique," he recalled. "The other astronauts are packing up, they look out the window, they laugh, they giggle, they do all the things that people do when they're relaxed. But the commander is never relaxed. You can't relax until the wheels come to a stop on the runway and its over." *Columbia* rolled to a stop on the 15,000-foot-long concrete runway in Florida, and Allen's first and only command came to an end. Despite the burdens and setbacks, he recalled the opportunity fondly. "It's fun to be king," Allen said.

Future kings may guide NASA spacecraft through the hazardous trip from Earth to Mars. If that crew and future pioneers are to survive on the hostile surface, they may be called upon to set up their own enclosed habitat. That may not be as easy as it sounds, if the experience of Biosphere 2 in Arizona is any example.

LESSONS FROM
biosphere 2

Sending human beings to Mars will involve severe risks. The crew will be cut off from Earth with little hope of rescue or quick resupply in case of a catastrophe. They will face the fiery entry into the thin atmosphere and have one chance at surviving touchdown to the hostile surface of the red planet. If something goes wrong, mission control and the rest of humanity may never know what happened to the astronauts. Even if everything goes perfectly, the crew will face endless months of isolation, just as American trailblazers did while taming the Old West.

The longest nonstop endurance flight in orbit was aboard the Russian space station *Mir*. Cosmonaut Valery Polyakov spent 438 days circling Earth, ending in 1995.[1] His flight was punctuated by the historic first rendezvous of *Mir* with Space Shuttle *Discovery*. Both vehicles edged up to within thirty feet of each other.[2]

Fourteen months cooped up inside *Mir* provided data on the impact of long-term space missions on crew members. In their own way, an oil tycoon and eight determined volunteers attempted something similar inside an airtight complex the size of three football fields. It was called Biosphere 2. Critics complained it was a publicity stunt, but proponents claimed it was a serious attempt to put a small slice of the world's ecosystem into a bottle and make it self-sustaining.[3]

For the family of Jane Poynter, it provided the comfort of knowing where she was, even if that meant being confined for two years inside an enclosure resembling a high-tech cathedral. "I think they were rather relieved," says Poynter. "I've lived in these crazy places over the years, and when you're out at sea, it's tough to communicate. So I think they were happy about it." The adventurer, researcher, and entrepreneur was invited to join seven other people to make the first long-term mission inside Biosphere 2 in 1991. She ran into Biosphere's organizers by accident while doing research in the Australian Outback on grassland that was overgrazed by sheep ranchers. From there, she found herself outside of Tucson, Arizona, where oil tycoon Ed Bass was preparing to spend $150 million of his own money to build a giant terrarium covered with over 6,000 panes of glass.[4]

Poynter grew up reading science fiction stories by Isaac Asimov and dreamed of being an astronaut. Having the chance to participate in Biosphere 2 was probably the closest she would ever come to flying in space. But when Poynter arrived, all there was to see was a barren patch of desert and a visitors' center. "When I arrived on the job the first day," she recalled, "there were earth diggers and huge machines with this infernal beep, beep, beep, going on throughout the night." Poynter's task was to manage the design of the agriculture section of the three-acre complex. Other specialists would be in charge of the health of the "Biospherians," as well the architecture and the mechanical systems. The organizers operated under the theory that the people building Biosphere 2 would do a better job if they were among the ones living inside.[5] "I was absolutely enthralled and fascinated by the concept of closing up and seeing if it would persist," Poynter said. That enthusiasm would change over the months.

The goal of the complex was to operate while sealed off from the outside world. Food, water, and even the air were supposed to be generated and recycled without resupply. Biosphere 2 included a miniature rain forest, an ocean with a coral reef, and land for growing the food supply that would be eaten by the eight inhabitants as well as the animals kept in the enclosure.[6] The crew members weren't supposed to use the pigs, chickens, and goats as sources of meat. Instead, the animals were part of the recycling system.[7] "Peanuts were the primary source of fat in our diet," says Poynter. "We couldn't eat the leftover peanut greens, which would have gone to waste, but the goats could." The result was an ongoing source of milk for the crew.

The water filtration system also utilized the plants growing inside the enclosure. Wastewater from the crew members was sent through a small marshland for partial purification. It wasn't fit to drink, but it was usable for irrigating the crops of sweet potatoes, carrots, beans, and rice. That process finished recycling the water for human consumption.[8] "The plants would soak up the treated wastewater, and what came out of their leaves was essentially purified water," says Poynter. "That went into the atmosphere, and we had heat exchangers that condensed the water out of the air, and that closed the cycle." The life-support system also created limitations. Farming within the compound had to be totally organic. No chemical fertilizers or pesticides could be used since they would end up in the water supply and poison the crew. Supporters claimed that what they were creating was a possible prototype for a base on the surface of Mars.[9]

The Mission Begins

On September 26, 1991, Poynter and her seven crewmates entered Biosphere 2 and locked themselves inside for an endurance mission lasting two years. The four men and four women wore matching blue jumpsuits as they waved good-bye to the outside world and shut the doors. The participants could send out blood samples for medical tests through a small airlock, but nothing from outside was supposed to go in.[10]

The complex's interior looked like a comfortable dormitory. Each crew member had an individual bedroom and office area. They shared bathroom facilities and a communal living room. Since the mission of Biosphere 2 occurred in the early 1990s, the Internet wasn't in wide use. There was a rudimentary system of e-mail within the complex, and telephones connected the crew to the outside world.[11] According to Poynter, the big problem area was the kitchen.

Members of the Biosphere crew took turns serving as chef for the team, and not all of them were gifted cooks. The self-sustaining nature of the mission meant that breakfast, lunch, and dinner had to come from ingredients grown inside the complex. Acreage was limited, so every square inch had to be used for producing crops. Things like herbs that could add interest to the day's menu were considered a luxury. Each cook got a tub of vegetables harvested from the farm, and it was his or her responsibility to prepare a day's worth of meals for everyone using those ingredients and nothing else.[12] "Cooking was very challenging," recalled Poynter. "There were no ready mixes or cardboard boxes to open in a panic so you could fix something if you messed up." The result was painfully frequent. The Biosphere crew would sit down to a watery soup that left them feeling constantly hungry. The day's workload included a lot of physical labor like tilling the soil, harvesting crops, or maintaining the machinery. The amount of calories in their diet wasn't enough to keep the Biospherians feeling full.[13] One of the few culinary breaks inside the complex was a pizza night, arranged by Poynter. "It took four months," she said. "We had to grow the wheat. Then get the goats pregnant to generate milk for the cheese." As the months wore on inside Biosphere, the subject of food grew into an obsession.

TV Dinners with a Twist

The lack of variety in the foods inside the complex left the crew members with barely enough energy to work, but little else. At the end of the day, they would gather in the living room to watch television.[14] That's where the deficiencies in the kitchen took another unusual turn. Jane Poynter says she and her crewmates became obsessed about the food featured in the programs. Everyone forgot the plotline. All that mattered was what the characters were eating and drinking. "We also indulged in what we called 'food fantasies,'" says Poynter. "We'd sit around and imagine in gory detail all the kinds we weren't eating like chocolate, and strawberries, and cappuccinos and red wine." Perhaps the worst example of this food obsession was when friends from the outside came to Tucson. Poynter arranged for them to have dinner at a local restaurant, but there was one condition. She supplied them with a videophone and asked that they keep it on during their meal, so she and the other Biosphere crew members could watch.[15]

Farm Tips by Long Distance

Telephone access did mean members of the Biosphere crew could contact people outside the confines of the complex. When questions arose about crops growing beneath the cavernous glass domes of Biosphere, Jane Poynter would place a call to South Bend, Indiana. That was home to Purdue University and its Center for Regenerative Life Support.[16] Director Cary Mitchell was one of the featured speakers at a conference for closed ecological systems at, but not inside, the Biosphere complex in Arizona. "Jane would call up, and we'd have great discussions about plant growth," says Mitchell. "It was fascinating." Mitchell had worked extensively with NASA to help the agency design a self-sustaining life-support system for outposts on the Moon or Mars.[17] However, it was clear that Biosphere wasn't getting a warm reception from the press. The project and its reclusive financier were the subject of unflattering headlines, some comparing the group to a cult.[18] Cary Mitchell knew from his association with NASA that the high-profile nature of the Arizona complex wasn't going to go over well with the space agency. "I think NASA kept Biosphere at arms' length," recalled Mitchell. "There was a lot of arm waving by the owners and those involved with it. NASA was leery of the kind of publicity that came out of it." The environment inside the facility depended on everything going smoothly. The plants needed to grow at a certain rate to utilize the carbon dioxide exhaled by the humans and animals in order to produce more oxygen. From Mitchell's perspective, things weren't going right for the people inside Biosphere 2. "The first winter they spent in there was the cloudiest in Arizona in years," he said. "The crops they put in there were growing slowly, so the carbon dioxide was building up." Mitchell didn't recall anything in Jane Poynter's voice that indicated a growing problem, but the rising levels of carbon dioxide and the falling amount of oxygen would prove to be a huge issue.

Four against Four, and Going Mad

The oxygen in the Biosphere was meant to be in the same amount as in Earth's atmosphere, typically 21 percent, with the remainder being other gases, mostly nitrogen. As the months dragged on inside the enclosure, the oxygen began to disappear. By the winter of 1992, levels has dropped to 14 percent, which was threatening the health of the crew.[19] Phone conversations between Jane Poynter and Cary Mitchell continued, but there was no clue in her voice that anything was wrong. "It wasn't like there was helium in the air," says Mitchell. "She wasn't sounding like Alvin the Chipmunk or anything." But for Poynter and her crewmates, the situation was becoming intolerable. The symptoms resembled high-altitude sickness suffered by mountain climbers. Usually, patients overcome this illness when their systems produce more red blood cells to carry more oxygen. Later analysis of the Biospherians showed their low-calorie diets didn't give them enough energy to generate the extra blood cells they needed.[20] The only alternative was to suffer the ill effects. "We were very tired all the time, there were mood swings, clinical depression, and sleep apnea," says Poynter. That's a medical condition where the patient temporarily stops breathing while asleep. "You gasp and it wakes you up," she said. "It does that over and over again, so you can't sleep." The affliction led to additional stress, and the eight members inside Biosphere split into two factions of four. Arguments began, and crew members started eating emergency rations to fight their hunger. Supplemental oxygen was later pumped in, which prompted even harsher news headlines questioning the viability of the project.[21] "I thought I was going insane," recalled Poynter.

On September 26, 1993, the eight crew members opened the doors and walked out. The two years they had spent inside were a blur. Almost immediately, Poynter, Cary Mitchell at Purdue, and others began to study what had happened. One concern was whether the astronauts on a long-term mission to Mars might suffer similar hardships compared to those inside Biosphere 2.

Compost and Concrete

The falling oxygen levels within the enclosure were a mystery while Poynter and her colleagues were inside. One member of the team had the job of monitoring how much carbon dioxide was in the air to ensure the safety of the atmosphere. The carbon dioxide remained constant throughout the crisis.[22] Later, a hypothesis emerged with a number of factors all combining to create the air problem. First, the compost used to fertilize the crops utilized more oxygen than expected. Also, the concrete used to build the complex absorbed some of the carbon dioxide, which deprived the plants of what they needed to produce oxygen. What was bad for the crops was bad for the humans.[23]

Compost and concrete were valid concerns to Cary Mitchell and his team at Purdue, but the psychological problems in Biosphere 2 were especially important. "NASA had really ignored the psychological aspects of long-term space flights," says Mitchell. "They figured if their crews had the 'right stuff' they'd be fine. But Biosphere proved that people are going to be people, and there are going to be interpersonal conflicts in that kind of isolation."

Even the number of crew members is believed to be a factor in the situation. Four men and four women were selected for the initial mission in the compound. Researchers at the Johnson Space Center declared that to be a mistake.[24] "Three wasn't a good number because you'd have two on one," says Poynter. "Any odd number would be fine, but eight was bad because you could split into two groups of four, and we had warring factions." Even though Biosphere covered more than three acres, members of the team felt they couldn't get away from each other to blow off steam. Modes of release like television and telephones to the outside world didn't appear to be enough. "Normally on the outside, you could go to a movie, or talk with friends and get in a better mood," says Poynter. "But in Biosphere you couldn't do that. And the same thing will happen on Mars." Despite all the problems and the mixed reviews in the press, the project still has its defenders who fiercely object to the image of Biosphere as some kind of joke. Proponents claim the structure achieved things that even NASA couldn't do, like maintain an airtight seal.[25] The International Space Station, for example, suffers from minor air leaks. "From an engineering perspective, one of the best things that came out was how they avoided the air leakage rate," said Cary Mitchell. "They built side chambers called 'lungs' where the air could go when Biosphere heated up during the day. It was a beautiful design." Since the heyday of the two long-duration missions in Biosphere, the owners once considered using the property for a housing development, and the complex was turned into a tourist attraction.

As NASA works to move ahead with its own plans for outposts on the Moon or Mars, the debate is whether to go with a bio-regenerative system like Biosphere 2 or with the mechanical systems being tested aboard the International Space Station.

ᴘʟᴏᴡɪɴɢ ᴛʜᴇ
"ᴄᴀᴄʜ ʏᴏ" ᴏɴ ᴍᴀʀꜱ

If NASA sends astronauts for long-duration missions to Mars, the crews will face dangers not seen by explorers since the pioneering of the western frontier. The U.S. space agency will likely have to undergo a change in its mind-set, which was forged in the early 1960s.

When President John Kennedy challenged America to make the Apollo lunar landings, NASA couldn't do it. The fledgling space program lacked the ability to send people into orbit around Earth, let alone travel to the Moon.[1] That goal was achieved with the landing of Apollo 11 in 1969. Designers of future trips to Mars say NASA will have to undergo additional growing pains to make it to the red planet. Astronauts will no longer be able to think of their missions as vacations. The agency will have to change from "trekkers" to "explorers." Trekkers, or backpackers, carry the supplies they need and head home when their provisions run out. Explorers, by contrast, live off the land.[2]

Every one of NASA's space missions up to now has included astronauts as backpackers. Their trips in space were limited by how much food, water, oxygen, and fuel the crews could carry with them. There's no greater example of this than the Space Shuttle. Sitting on the launchpad, most of the spacecraft's weight is made of up fuel. There's just enough inside the external fuel tank and twin booster rockets to achieve orbit. Ninety-five percent of the liquid propellant in the tank is burned just to finish the trip to space. The fuel margin is so tight that sensors inside the tank are designed to shut off the engines if the shuttle runs out of gas prematurely. In that case, the motors' turbo pumps might run wildly out of control and explode.[3] The issue of supplies and their impact on the length of a space mission was serious enough for the Russian space program to redesign its Salyut space stations in the 1970s. The one docking hatch on the earlier models was replaced with two, one in the front to receive Soyuz ships carrying crew members, and a second at the back primarily for resupply ships from the ground. The change made longer missions possible. More food and fuel could be sent up, and fresh Soyuz craft could be delivered to replace aging capsules for the final return trip to Earth.[4] Like vacationers, "roughing it" might be as simple as waiting for room service. On Mars, routine cargo trips will be difficult, if not impossible, due to the distance and expense.

Proponents of a self-sufficient Mars base look to Antarctic explorer Ernest Shackleton for inspiration. When he ventured to the South Pole in 1914, he didn't bring supplies for the two-year trip across the continent. Instead, Shackleton brought tools and weapons so his crew could survive by using whatever they could find around them. Astronauts on Mars will likely be called upon to subsist in a similar fashion.[5]

Asparagus on Mars

If you've ever wondered why French champagne has a "mineral" flavor to it, connoisseurs say it's due to the chalky soil where the grapes are grown in the Champagne region of France. The taste of the fruit is a reflection of where the vines are planted. Winemakers call the impact of these growing conditions *terroir*.[6] Scientists working with NASA are already speculating on what vegetables grown on Mars might be like, and whether the flavor would be "otherworldly" at all.

Speculation on this subject followed the landing of the *Mars Phoenix* spacecraft, which touched down near the planet's North Pole in May 2008. Mission managers immediately ordered the boxy, three-legged lander to use its robotic arm to scrape up soil to look for signs of ice. Within one month, a test to determine the chemical content of the dirt on Mars prompted some unusual comments to reporters. The experiment was called the Microscopy, Electrochemistry, and Conductivity Analyzer, or MECA for short. The spacecraft took soil samples with the robot arm and dumped them into a miniature oven on the vehicle. The samples were then "cooked" and scanned for signs of carbon, oxygen, and minerals. During a press conference, team leader Samuel Kounaves of Tufts University proclaimed that the soil on Mars wasn't chemically different from what's found on Earth. He went on to state that asparagus, turnips, or green beans may grow just fine in that type of soil. "We basically have found what appear to be the requirements, the nutrients, to support life past, present, and future," said Kounaves. "The sort of soil you have there is the type of soil you'd have in your backyard." Crew members on Mars could sustain themselves by tending rows of sweet potatoes or by maintaining mechanical systems that do many of the same tasks. There are advocates on both sides.[7]

The Dollars and Cents of Survival

During the height of the Space Shuttle program, NASA operated under a "rule of thumb" regarding the cost of putting a pound of anything into low Earth orbit. Whether it was food, water, equipment, or people, the price was set at about $3,000 per pound. Estimates vary at the expense when astronauts return to the Moon, but the space agency puts the figure at about $15,000.[8] Cary Mitchell's team at Purdue University, which is working on bio-regenerative systems for NASA, puts the cost of sending a pound of cargo to Mars at a whopping $140,000. NASA estimates that an average astronaut will require 70 pounds of food, water, and oxygen each day to survive. A crew of four, on an estimated two-year mission to Mars, would need over 400,000 pounds of supplies.[9] If those provisions were carried in bulk on the spacecraft, the launch price for NASA could be $28 billion. Proponents of recycling say this astronomical cost makes reusing smaller amounts of supplies a fiscal necessity.

Gears and Bolts versus Little Watering Cans

Supporters of agriculture on a Mars base believe plants can filter waste water and generate oxygen, as well as provide a source of food. They admit, however, the concept has an image problem with the general public. The 1972 motion picture *Silent Running* featured actor Bruce Dern as an astronaut aboard a space freighter. The vehicle is equipped with cavernous transparent domes with the last of Earth's trees and

This is an artist's conception of what a greenhouse module on the surface of Mars might look like. Experts disagree as to whether a glass or plastic dome is the best idea. Photo courtesy of NASA.

plants inside. Small, two-legged robots named Huey, Dewey, and Louie used little water cans to tend the gardens. Artist conceptions for NASA followed this idea with clear domes on the Moon or Mars.

This vision may work in a movie, but proponents of greenhouses on Mars think it's impractical. Gioia Massa, a researcher with Cary Mitchell's team at Purdue University, specializes in the kinds of artificial lighting space agriculture might require. She believes the public's perception for growing things in space was skewed by pop culture like *Silent Running* and by Biosphere 2. "It was very much like a garden of Eden in that they had these open soaring spaces and diversity of biomes," says Massa of Biosphere 2. "For a Moon or Mars mission, we're not going to have that." If NASA sets up greenhouses on the Moon or Mars, planners will have to adjust to conditions as they are, not as Hollywood would have them. First of all, advocates say don't try to use domes covered with glass or plastic. Cary Mitchell blames the soil, or regolith, at both locations. "The lunar regolith is very sharp and interlocking," he says. "It clings to plastic or glass surfaces and would likely cover over a greenhouse." The same issue is expected on Mars, where dust storms are expected to send clouds of red dust blowing around the surface. Safety is another consideration. The thin atmosphere around Mars doesn't screen out harmful radiation from the Sun, and incoming meteorites aren't expected to burn up as thoroughly as they do on Earth. That could make domes more vulnerable to radiation exposure or to cracks and punctures from falling space rocks. Just as astronauts will have to wear spacesuits to move around outside, plants on the Moon or Mars will have to be protected from hostile growing conditions, such as the different amounts of sunlight and gravity.[10]

Farmers on Earth produce vegetation that's used to a certain amount of sunshine and darkness. If NASA sends crews to live at one of the Moon's polar regions, light from the Sun is expected to be almost constant. Teams on Mars will face a different challenge, since daylight there is dimmer due to the planet's greater distance from the Sun. "One of the things that plants do is that they have a bunch of signals," says Massa. "Light is a signal for growth. Water flow is a signal. Plants will grow toward sources of water. Gravity is one basic signal. But in the absence of these, they can key in on one of the others." That's another reason why the Purdue team believes that surface domes for growing crops won't work. "We think that plant growth isn't going to be in a surface greenhouse," says Mitchell. "It's going to be underground, and electrical lighting will be used." Their concept would be to put cylindrical modules in caves or lava tubes on Mars to protect the plants from radiation. The inside of each compartment would be lined with plant trays and equipped with artificial lighting overhead. The modules could be buried by astronauts using heavy equipment or by robotic hardware working autonomously before the crew arrives from Earth. "Volume is going to be the biggest constraint," says Massa. "We're going to have to grow as many plants as we can in small spaces with the plants packed in tightly." Artificial lighting is intended to promote growth even with Mars's gravity, which is only one-third as strong as on Earth. Instead of rows of bulbs or fluorescent tubes, Purdue envisions long strips of LED lights inside the greenhouse modules, similar to the emergency floor lighting of a passenger jet.

What to Grow?

NASA put together a list of candidate fruits and vegetables to be grown in space greenhouses. The goal was to design a nutritious vegetarian diet since animal by-products may not be available. Biosphere 2 included goats, pigs, and chickens as part of its self-sustaining design. NASA, however, appears unlikely to turn its Mars mission into a high-tech "Noah's ark," so the astronauts will likely eat only what they can grow, assuming the agency selects a bio-regenerative system. The list of crops includes wheat, rice, soybeans, sweet potatoes, white potatoes, and peanuts.[11] "You need sources of proteins, you need sources of fats," says Mitchell. "And you need some sources of starches and put that together for a balanced diet." To avoid the problem of the constant hunger faced by the crew of Biosphere 2, mission planners are also thinking of "salad" crops, which could add flavoring and texture, such as tomatoes, lettuces, spinach, strawberries, and herbs. "The staple basis of a vegetarian diet is a ratio of cereal grains to legume seeds in a certain proportion of proteins," says Mitchell. The point is to approach the levels of proteins people normally get from meat, milk, and eggs.

NASA commissioned studies at Purdue and elsewhere to examine gardening on long-term space missions. It was a program called the Controlled Ecological Life Support System, or CELSS for short.[12] It operated between 1990 and 1995, and prompted several warning flags. Number one, how would the astronauts survive a major crop failure with help from Earth months away? Some mission planners want to include a large supply of emergency nutrition food bars. Their flavor is compared to cardboard, but supporters say they could keep astronauts alive through an agricultural crisis.[13] Designers at Purdue doubt that their greenhouse system would suffer a permanent breakdown. Their confidence comes from the way they define a crop failure. "If the conditions are correct, the plants won't fail," says Massa. "It's usually some aspect of the life-support system controls, something that's machine-made or man-made." Likely culprits could be a problem with the lighting or water-flow system. But Purdue believes it's nothing from which the astronauts couldn't bounce back. "If you give the plants what they need, they'll grow," says Massa. Advocates of bio-regenerative life support on Mars also say there are psychological benefits from space farming that NASA needs to consider. This became apparent when NASA tested space greenhouses on the same continent where it set up a prototype of its inflatable crew compartment.

NASA used McMurdo Station on the western coast of Antarctica to test its air-filled Moon base module. The New Amundsen-Scott Station is located northwest of McMurdo, and that's where NASA set up equipment to test the viability of growing plants on the Moon or Mars. Racks of plants were set up with artificial lighting, and station personnel were assigned to tend the crops.[14] One of the more interesting results had nothing to do with how the plants responded to conditions inside the growth facility. Rather, investigators noted how the people at the New Amundsen-Scott Station responded to the plants. Farming was considered a welcome break from life in the harsh conditions of Antarctica. "Some of them looked forward to tending the plants," says Massa. "They would find people sleeping in the greenhouse to be around green living things." University researchers have long speculated that astronauts living on Mars, and waking up to the same rusty-colored landscape every day, would need a comforting reminder of life on Earth.[15]

Along with coping with crop failures, NASA also studied the amount of manpower needed to keep a proposed Mars greenhouse going. The International Space Station was criticized for not accomplishing much science with the crew limited to two or three astronauts and cosmonauts. The public perception was that the smaller crews spent most of their time maintaining broken equipment on the facility rather than doing the experiments NASA promised.[16] Future news coverage of astronauts tending wheat and soybeans on a multibillion-dollar Mars flight is something the agency would likely want to avoid.

The Cost Beyond Dollars

NASA coined a phrase for the "costs" of launching and maintaining any life-support system, whether biological or mechanical. It's called "equivalent system mass." It goes back to the hard-earned lesson at NASA that nothing involving space travel is simple, and how things work on Earth doesn't count.

If you want to grow some tomatoes in your backyard instead of buying them at the supermarket, it's relatively easy. Depending on the weather in your region of the country, the process could involve little more than buying seeds, some fertilizer, and a watering can, and setting aside a patch of soil. Even city dwellers could do something similar in a window box if they chose. Earthlings have it easy that way. The planet has an atmosphere that supplies oxygen and screens out deadly ultraviolet radiation from the Sun. There's water readily available from the tap. Also, gravity keeps things from floating around uncontrollably. And if your plants die, you can pick a restaurant and eat out. In space, none of this can be taken for granted. Astronauts on the Moon or Mars will be living in a hostile and alien environment, and everything including water, power, and infrastructure will have to be brought from Earth or generated by the crew.

NASA's process for calculating equivalent system mass includes how much a given life-support system weighs during liftoff and on the trip to the Moon or Mars. Other factors include how difficult it is to move and set up, how much electricity it requires, and whether a pressurized compartment with its own air supply is needed. Every answer includes a trade-off. An unpressurized crew module, for example, would require astronauts to wear spacesuits to maintain it. Spacewalks are costly in time and effort. This scrutiny is used in judging bioregenerative systems like greenhouses, or life-support designs that rely on high technology.[17]

Breathing Moon Dust

Years before the crew aboard the International Space Station began experimenting with equipment to recycle their own waste water, K. R. Sridhar was looking for a similar way to make his mark at NASA. After completing his doctorate degree in mechanical engineering at the University of Arizona at Tucson, he wanted to work on space-related projects. He came to a quick realization about the way NASA selects missions. "It's a political thing," he said. "We technologists don't control that destiny. It's left up to somebody else." That drove Sridhar to ask himself the fundamental question, what would any long-term space trip need? He decided it was oxygen for the crew to breathe. "I wanted to find a common technology to provide oxygen that applies to the Moon, or Mars, or asteroids," says Sridhar. "It doesn't matter where NASA goes, I'll still be in business." The agency sent crews to the Moon in the 1960s, and into orbit on the Space Shuttle beginning in the 1980s, with oxygen kept in tanks. Future inhabitants of a Mars base could breathe air that way, or they could create it.[18]

The thin atmosphere on Mars is 95 percent carbon dioxide, so a mechanical system could theoretically be used to extract the oxygen. For Sridhar, the Moon was the bigger challenge, since it has virtually no atmosphere at all. He focused on how to pull oxygen from the most abundant resource on the Moon. It was the soil where the Apollo astronauts left their boot prints from 1969 to 1972. Specifically, he wanted to harvest oxygen from an oxide-rich type of Moon dirt called ilbanite. From Sridhar's viewpoint, it was the equivalent of creating fire or the wheel, assuming it could be done.[19]

Selling NASA on the Idea

Sridhar's proposal to the space agency was to create a process that combined ilbanite with carbon. The mixture would then be heated until it produced carbon monoxide. From there, the oxygen could be extracted for the astronauts to use.

Simple, right?

In 1992, NASA supported the idea and funded Sridhar's research with the goal of creating a small amount of oxygen using the ilbanite extraction process in the laboratory. "It's like making steel," he told the agency. "Iron ore is mostly iron oxide. You combine it with carbon in a blast furnace, and you get steel." The problem was generating enough heat to cause the chemical reaction without draining the power supply of a Moon or Mars base. Sridhar compares the situation to that of a rabbit trying to survive in the wild. "If you're a rabbit looking for food," he explains, "and for every 100 calories of food you eat, you spend 200 calories looking for it, you're going to die." Sridhar's answer is to utilize the Sun for the heat needed for the process. A solar collector, called a concentrator, could focus sunlight on the ilbanite and carbon mixture to provide the heat needed to generate carbon monoxide, and later oxygen.

Once this process was perfected, Sridhar believes, doing the same thing on Mars would be easier since the astronauts could skip the step of extracting oxygen from soil and just "suck" it out of the Martian atmosphere. That process is called solid oxide electrolysis. The design calls for a blower to pull Martian air across a ceramic surface that separates oxygen molecules from carbon dioxide with an electrical current. A similar system was developed by the Russian space program to extract oxygen and hydrogen from water aboard the space station *Mir*. "Think of it as an electro-chemical sieve," says Sridhar. Applications might include units the size of a fanny pack, which astronauts could strap on to their spacesuits. "As long as they have power for their electro-chemical sieve," he contends, "they can be gone as long as they like, with no need for an oxygen tank." A larger version of the device, perhaps the size of a refrigerator, could provide breathable air for an entire crew compartment.

NASA's Johnson Space Center supplied Sridhar with ilbanite similar to soil samples collected by the Apollo astronauts to test his idea. The result was a few puffs of oxygen created in the lab. That success led to a follow-up agreement with NASA to generate several liters of oxygen using the same process.

Brimming with confidence in their system, Sridhar's team at the University of Arizona submitted a plan to NASA to test the technology on an upcoming spacecraft called *Mars Surveyor 2001*.

The ambitious Mars mission was supposed to carry a miniature rover called *Athena*, which would crawl around on the surface. But perhaps the highest-profile part of the lander was a small rocket designed to deliver Mars soil samples to Earth. The missile was intended to address a major complaint by designers of a manned mission. They could dream up any technology they wanted that utilized Mars soil, but until they could work with actual samples from the planet, it was all theoretical.[20]

Even as Sridhar's plan for mechanical life support gained steam, the team at Purdue University hoped the return rocket might prove that their plan for agriculture on Mars had merit. The soil samples would be useful even if they drove planning in another direction. "You might find that you can't grow plants in the Mars regolith," says Cary Mitchell. "There might be something in the soil that's toxic to the plants or the astronauts." While researchers in Indiana hoped for a chance to experiment with Mars dirt, Sridhar wanted *Mars Surveyor* to prove that his hardware to generate oxygen from the thin and unwelcoming atmosphere of the planet could work.

NASA granted Sridhar and the University of Arizona a $20 million contract to develop an experiment for *Mars Surveyor* that could be no bigger than a microwave oven. The package was called the Mars In-situ Propellent, or MIP, experiment. The hardware would take air from Mars and extract the oxygen, which could eventually lead to a source of breathable air for astronauts or propellent for the return trip.

Then NASA's plans went sour.

Mars Surveyor was scheduled to launch after the ill-fated flight of the *Mars Polar Lander* in 1999. That spacecraft was believed to have crash-landed on the red planet due to human error. While NASA tended the wounds to its public credibility, the agency canceled *Mars Surveyor*'s mission. Sridhar's hopes to test his plan to extract oxygen from the Martian atmosphere were indefinitely grounded. The hardware from *Mars Surveyor* would find later life when it was rebuilt into NASA's *Mars Phoenix* lander. Sridhar's experiment wasn't included, and the unit sits unused in a warehouse.[21]

The disappointment over *Mars Surveyor* didn't quell the debate over mechanical life-support equipment versus plant-based bio-regenerative systems. Sridhar contends the advantage to his plan is its reliability. "When failure is not an option, and you can't bring people in and out, and it's a life critical need, bio-systems can be finicky," he says. "One virus or something could do a lot of damage." Inefficiency is another complaint. Every pound used in launching, building, or maintaining a Mars base carries financial and technological costs. Critics of greenhouses say they produce mostly unusable waste products like leaves and stems, as opposed to things the crew can eat. Even using crops to purify water or air, opponents contend, carries the danger of the complexity of the biological process. "Chemical or mechanical systems for life support can be built with redundancy and are predictable," says Sridhar. "We don't understand the biology of plants as a science as much as we understand chemical reactions." Also, there's considerable experience with chemical life support on NASA spacecraft, as well as earthbound applications like submarines, where crews live and work in an enclosed environment.

Proponents of bio-regenerative systems point to one major flaw in chemical life-support systems. You may be able to manufacture oxygen, but you can't manufacture food. Short-term missions to Mars may be able to function with provisions brought from Earth, but longer expeditions will have to be self-sufficient. "We've been having this tug of war with NASA for years," says Mitchell. "You have to have a balance between the two systems." Purdue contends that chemical and biological life-support plans don't have to be at odds with each other. Assuming NASA continues to blend ideas for long-term space missions, such as using both hard-shell and inflatable crew modules rather than favoring one over the other, a combination of mechanical and plant-based systems might eventually be chosen. Crews might experiment with test beds of plants during initial Mars missions while surviving on prepackaged food brought from Earth.

Along with how to sustain the crew, another decision NASA and the White House will have to make is whether to go alone or invite the same kind of foreign participation that made the International Space Station possible.

should nasa go it alone?

The debate over whether the United States should invite foreign participation in its proposed manned Mars initiative has dragged on for years. Discussion with foreign governments and NASA has been publicly friendly, but doubts remain within the U.S. space agency. International cooperation looks good, but it can complicate space missions as NASA moves from short trips, similar to vacations, to longer and more dangerous flights. During the hazardous trek to the red planet, some wonder if diplomacy is something NASA can afford. At one point, national security was an issue.

During the era of détente, there were complaints of NASA getting a little too cozy with the Soviet Union during the planning of the Apollo-Soyuz Test Project in 1975. The mission was the "final hurrah" for the American program to land men on the Moon. Three U.S. astronauts—Mercury pioneer Deke Slayton, veteran Apollo commander Tom Stafford, and rookie Vance Brand—blasted off aboard the last NASA Moon capsule on July 15, 1975. The same day, cosmonauts Alexei Leonov and Valeri Kubasov headed to orbit aboard a Soviet Soyuz spacecraft, which launched from Kazakhstan.

Their goal was to rendezvous and dock in an act of American/Russian cooperation. However, each vehicle's crew cabin operated at a different air pressure. The NASA command module used a pure oxygen atmosphere at five pounds per square inch. By contrast, the Soviet Soyuz used a mostly oxygen atmosphere at normal air pressure on Earth, about fifteen pounds per square inch. The only way the two vehicles could dock, and transfer people back and forth, was through

The Soyuz spacecraft is the workhorse of the Russian space program. First launched in 1967, the crew capsule later became a mainstay aboard the International Space Station, to carry astronauts and cosmonauts to and from the orbiting complex. It also was the emergency lifeboat in case of an evacuation. Courtesy of NASA.

an airlock.[1] NASA and Moscow designed and built a compartment with hatches on either side. It was called the "docking adapter." The hardware would launch aboard the same rocket as the Apollo craft, inside the same compartment that previously held the lunar modules that carried astronauts to the Moon.[2]

Once Stafford, Slayton, and Brand achieved orbit, they pulled their capsule away from the rocket, leaving one end of the new airlock compartment exposed. The Apollo craft docked its nose to the module, and then backed out with the airlock attached. The Russian Soyuz would later link up with the compartment so the crew members could conduct joint activities.[3]

Critics of the project complained that it gave the Soviets access to U.S. space technology. Moscow, it was pointed out, showed interest in a joint docking mission with NASA only after cosmonauts tried unsuccessfully to dock Soyuz vehicles together without foreign help.[4] The misgivings over working with the Soviets even led to a rumor within the Central Intelligence Agency that one of the Soyuz crew members, either Leonov or Kubasov, was a KGB agent sent to spy on the Americans.[5]

Proponents of Apollo-Soyuz point to the docking technology that came from working with the Russians. Instead of the older docking mechanisms that used a metal probe and a socket, the joint mission led to a new type of hardware with metal petals that came together like someone interlacing the fingers of his or her hands. This new idea was later used on the Russian space station *Mir* as well as the International Space Station. Supporters asserted both sides benefited through the relationship.[6]

Working with former Cold War foes is one thing, but even space cooperation with longtime allies wasn't without dissension. The European Space Agency (ESA) currently includes eighteen member nations ranging from Great Britain to Greece. ESA has launched its own probes to Mars, like the failed *Beagle-2* lander, and also joined with NASA on spacecraft like *Cassini*, to explore Saturn, and *Ulysses*, which studied the sun. Perhaps the highest-profile example is the seventeen-ton Spacelab, which Europe developed to fly in the cargo bay of NASA's Space Shuttles. The school bus–size lab carried experiments on missions dedicated to Germany and Japan, among others. There are also complaints among American scientists that when astronauts from Belgium, Italy, or France flew on the Shuttle, a U.S. scientist lost a chance to fly on a spacecraft funded by American taxpayers.[7]

Former astronaut Ken Reightler had heard the arguments for and against foreign participation and understood them. He admits he had his doubts along the way as well. His first Space Shuttle flight delivered the 13,000-pound Upper Atmosphere Research Satellite to orbit in 1991. The boxy satellite, laden with science equipment and a single solar electricity panel sticking out to one side, orbited Earth almost 80,000 times before being decommissioned in 2005. Its job was to sniff the atmosphere for changes in the ozone layer, among other things.[8] Once Reightler's delivery job was over, he was asked to take on a new responsibility. Among his other duties, Reightler was named deputy of a NASA task group to implement a mandate from newly elected president Bill Clinton. The White House wanted astronauts to fly on Russian space missions, and cosmonauts were to be invited to join at least one Space Shuttle crew. Reightler had heard the complaint that working with Russia was like "giving away the crown jewels" of NASA's hard-earned space technology.[9]

The International Space Station, which included hardware and personnel from the United States, Russia, Europe, Japan, and Canada, served as a symbol of how nations would work together in space. For Reightler, it also hammered home the fact that cooperation in orbit took more than stitching a foreign flag onto the sleeve of a U.S. spacesuit. "It's hard enough to do all the technical stuff without all the cultural differences, along with language, and all the logistics," he recalled of the effort to build the space station. "The argument could be made that it might simplify things if we go it alone." A manned trip to Mars could be one example of that.

One lesson NASA learned in working with the Russians was something simple, but critical. It was how either agency measured the amount of radiation faced by the crew members in orbit. NASA's detectors and the ones from Russia gave different readings. The solution was to put detectors from both countries in the same spot on the Space Shuttle and compare notes on what the devices said.[10] It was interpersonal work like this that changed Reightler's outlook on foreign cooperation in space.

A group of scientists meeting at Stanford University in 2006 needed no convincing. The delegates resoundingly voiced their support of asking many nations to share the financial costs and human risks of a voyage to the red planet. Scott Hubbard, who led the discussion, once sat on the board that investigated the tragic loss of Space Shuttle *Columbia* in 2003. He compares NASA's current mind-set on sending astronauts to Mars to building a highway. The United States, he contends, would be asked to build the infrastructure, with foreign space agencies contributing an off-ramp or two. "We are urging the new administration," said Hubbard, referring to the Obama White House, "to see if there is some broader international cooperation that could add this capability without putting the entire price tag on the back of the U.S."[11] Observers of the evolution of foreign cooperation in space believe there are specific things that other nations could contribute to NASA's efforts to visit Mars. It all boils down to hardware and relationships.

What Europe Might Bring to the Table

Just as NASA's capabilities have evolved over the years, so have those of ESA. Visitors to the Smithsonian Institution's cavernous Udvar-Hazy Air and Space Museum in Virginia can see examples of that evolution side-by-side. NASA's Space Shuttle *Enterprise*, which made unpowered glide tests in the mid-1970s, sits on display along with one of ESA's Spacelab modules. ESA's portable lab module paved the way for the more ambitious *Columbus* laboratory, which was attached to the International Space Station.[12] ESA astronaut Hans Schlegel helped deliver *Columbus*, and his job description has changed along the way as well. The veteran space flier took his rookie flight aboard Space Shuttle *Discovery* in 1993 on STS-55. The mission included the second Spacelab compartment dedicated to experiments from Germany. Schlegel and his fellow countryman Ulrich Walter worked alongside NASA astronauts Steve Nagel, Charlie Precourt, Tom Henricks, Bernard Harris, and Jerry Ross to do eighty-eight science studies during the mission.[13]

Fifteen years later, Schlegel was a member of the crew of STS-122, preparing to connect the new *Columbus* lab to the International Space Station. He would be joined by astronauts Steven Frick, Dan Tani, Alan Poindexter, Stanley Love, Leland Melvin, and Rex Walheim to snap the module into place. Instead of the simpler Spacelab, which was designed for limited exposure to the environment of space from inside the Shuttle's cargo bay, the more advanced *Columbus* module was built to stay in space permanently as part of the orbiting station.[14] Schlegel's job was to perform spacewalks to help snap the new compartment in place, next door to the U.S.-built *Destiny* laboratory. "It's a scientific lab, like the American lab," says Schlegel. "But the American lab has a lot of functions as well, so it doesn't concentrate on science, but *Columbus* does." The barrel-shaped *Columbus*, filled with experiment racks and laptop computers, was different from what ESA originally intended.

Columbus was first designed as a free-flying laboratory that would orbit Earth separately from the main space station. Budget cuts following the reunification of West and East Germany are blamed for killing that plan for *Columbus* and prompting the pressurized lab module now attached to the International Space Station complex.[15] Even the scaled-back European compartment had to be fit into the maze of compartments on the station and subsist on reduced power until all of the solar electricity panels could be successfully installed. "Everybody knew what we signed up for when we joined the International Space Station," says Schlegel. "Many things have to fit into each other. Every mission has to be successfully completed before the next one." The German astronaut's experience with foreign cooperation in space also extends to work between Europe and Russia.

It was fourteen years between Schlegel's 1993 mission on the German Spacelab mission aboard the Space Shuttle and his preparation for the flight to deliver *Columbus*. Between those trips to orbit, he went to Moscow to train for a mission aboard the Russian space station *Mir*. The objective was to get ready to leave the reasonably spacious crew cabin of the Shuttle and to be shoehorned into the smaller Soyuz capsule.[16] For Schlegel, it meant Russian-language lessons and the chance to live and work with the people there. "That was a tremendous effort and also a tremendous benefit to me," recalled Schlegel. "The main thing was I learned about my own prejudices." All in all, it didn't seem like fourteen years to him. It could be about that long before astronauts settle back on the Moon to prepare for the more dangerous trip to Mars. Officially, NASA is looking for ways for other nations to take part.

Europe demonstrated its ability to build space probes, to launch payloads with its Ariane rockets, and to build a crew compartment like *Columbus*. When it comes to providing hardware for a Mars mission, Kathy Laurini is NASA's point person. Her official title is Associate Manager for Lunar Formulation within the Constellation program to explore the Moon and Mars. Part of Laurini's job is to look for what other nations might "contribute" to the effort of returning to the Moon. Mars-related ambitions come later. "We're talking about what role Europe could play," Laurini says about a joint Moon trip, "whether it's a cargo carrier landing vehicle, or some other element, like a habitat module or a lunar lander." Extracting lunar resources like water or minerals could be another avenue for foreign participation.

For Laurini, the idea of foreign participation is a family affair. Her husband, Daniele, was named manager of the Italian-built cupola for the space station, which is a multiwindowed observation dome. Once the module is attached to the space station, crew members would use a mechanical hand crank to open the shutters on the cupola for the complex's most unobstructed view out into space.[17] Still, the Laurinis have worked to keep office talk away from the dinner table for the sake of their children. After clocking in for work each day, Kathy Laurini handled the issues raised by asking a number of countries to participate in NASA's proposed Moon program. These are issues she's faced before.

Prior to her assignment with the Constellation program, Laurini was in charge of integrating another European contribution to the International Space Station. It was an unmanned cargo craft called the Automated Transfer Vehicle (ATV). The very first one would be dubbed *Jules Verne* after the French science fiction writer of *20,000 Leagues Under the Sea*. The carrier would launch aboard an Ariane 5 rocket, which was the biggest booster Europe had.[18] It would improve on the capability of the Russian Progress freighter, which was a robotic version of the manned Soyuz craft. *Jules Verne* would carry three times more supplies than its Russian counterpart. The European cargo ship was built to launch from Earth, chase the space station, rendezvous with it, and then dock with a load of supplies on board. The European ATV craft would also act as a temporary crew compartment for six months before being discarded for a fiery reentry into Earth's atmosphere. It was a leap forward for ESA in terms of technical and diplomatic complexity. Instead of building on the European *Columbus* lab, the ATV traces its lineage to cargo modules developed by Italy that rode to orbit aboard the Space Shuttle. The Multi Purpose Logistics Modules, or MPLMs, would go by the more familiar names of Italian artists like Donatello and Leonardo.[19]

The *Jules Verne* craft closely resembled most of the silvery crew compartments on the station, except for a pair of blue-colored solar panels forming an "X" near the nose. The finished vehicle was the size of a double-decker tour bus you might find in London. The process of successfully docking the European craft on autopilot would be more complex than just pulling up and parking.[20]

The International Space Station is comprised of a cluster of crew compartments at the center of a long, skinny metal framework with two sets of solar electricity panels on either end. The ATV's target was the back docking port on the Russian-built *Zvezda* crew cabin. So Laurini's job to integrate the vehicle was compounded by getting Moscow,

Europe, and NASA cooperating all at the same time. "We really had some difficulties with interface issues," recalled Laurini. That meant ensuring that the ATV being built on the ground would connect perfectly with the *Zvezda* module, which was already in orbit. "We had to set up weekly teleconferences to get those details ironed out. It was an extraordinary effort," she says. Those details may have been tedious, but the ATV couldn't work without them.

To do its job, *Jules Verne* had to unload part of its cargo automatically. Crew members could float in and carry out items stored in the forward cargo compartment, but things like water and fuel had to be pumped in. Like the Russian Progress ships, the ATV would connect air and fluid lines between itself and the station as it docked with the complex. "How do you transfer the fuel? How does *Zvezda* transfer oxygen back into the ATV for the crew to breathe?" asked Laurini at the time. "If something goes wrong inside *Jules Verne*, how does it tell the station about it?" Even more to the point, the maiden flight of any spacecraft carries safety concerns because ground controllers and the astronauts inside the space station have no practical experience with it. The ATV would be the first totally new spacecraft to visit the space station in a decade.

Jules Verne launched successfully on March 9, 2008. On April 3, the vehicle was on its final approach with the crew of the International Space Station hovering over a red button that would abort the docking if something went wrong.[21] The outpost's first female commander, Peggy Whitson, along with crew members Yuri Malenchenko and Garrett Reisman, watched the ATV close the gap. "Right now it can be clearly seen, lit by the Sun," Malenchenko told ground controllers. *Jules Verne* crept toward its target at the aft end of the *Zvezda* module and locked itself in place without incident. During its six-month mission attached to the International Space Station, the interior of the *Jules Verne* spacecraft became a favorite among the station crew for sleeping and bathing. The vehicle's onboard thrusters were also used to raise the altitude of the station.[22] For Laurini, the various tasks performed by the ATV demonstrated the possible usefulness of foreign participation in a manned trip to the Moon or Mars. "It's what we call dissimilar redundancy," she explained. "Whether it's a different way to process urine or oxygen, so if you have different systems with different approaches from different countries, I think you're more likely to have more robustness. It provides more reliability." While some people are working out the technical issues of cooperation in space, others consider personal relationships to be the starting point of success.

Election Day and Thanksgiving

When astronaut Ken Reightler and veteran Space Shuttle commander Richard Covey were asked to set up a training program to put a Russian cosmonaut onto a Shuttle mission, they joked about being assigned to the flight themselves. Each man was acquainted with military training that utilized foreign officers. Reightler had attended the navy test pilot school, which includes permanent staff members from the British Royal Navy and Air Force. The thinking was to pattern NASA's efforts to get cosmonauts ready to go on a Shuttle flight on that military example.[23] Still, there were doubts. "I think initially there were some mixed feelings on whether this was a good thing or a bad thing," says Reightler. "The worry was it was going to distract us [NASA] from our primary mission like redesigning the International Space Station." At that time, NASA was working with the original station proposal called *Freedom*, championed by the Reagan administration and the George H. W. Bush administration. Once Democrat Bill Clinton was elected to the White House, the agency was ordered to incorporate Russia into the space station project along with Europe and Japan. Cosmonauts on the Shuttle were judged to be a possible complication.[24]

Still, Reightler and Covey pressed ahead with work to accept at least two Russian space veterans into the astronaut corps. They quickly learned that the process was more complicated than simply putting out the welcome mat.

NASA's Johnson Space Center is a secured federal installation, where personnel have to pass through checkpoints with photo badges. Even experienced space crew members from Europe can't just walk from place to place as they choose. One example is astronaut Claude Nicollier of Switzerland. He was already a veteran Space Shuttle crew member, having flown the first mission to carry the Italian tethered satellite

European Space Agency astronaut Claude Nicollier flew aboard two missions to deploy the Italian tethered satellite. He also took part in the flight of Space Shuttle *Endeavour* to restore the fuzzy vision of the Hubble Space Telescope. Courtesy of NASA.

experiment. However, the fact that he wasn't an American citizen created limitations. "He lived in Houston for years," recalled Reightler. "But still he had to be escorted to the simulators to train for his flights."

Along with security concerns for the incoming cosmonauts, basic living conditions had to be considered as well. NASA decided that apartments would be good accommodations for their Russian guests, since that's how they lived in Moscow. It also eliminated the problem of mowing the lawn in front of a house. The University of Houston was enlisted to help out with language courses, since NASA wasn't sure how fluent the cosmonauts would be with English.[25]

Top managers had long decided the first cosmonaut crew member would fly aboard Space Shuttle *Discovery* in late 1993 or early 1994. Its mission was designated STS-60. Its primary payload would be the first flight of the Wake Shield Facility. The twelve-foot-wide stainless steel disc would be grasped by the Shuttle's robotic arm, dangled over the side, and set adrift to orbit Earth with the blunt edge facing forward. The point was to create a "wake," which would funnel particles of molecular oxygen in the thin upper atmosphere where the Shuttle operates. Scientists hoped the oxygen would stream around the backside of the disc, creating an ultraclean environment for experiments on the Wake Shield, which might lead to things like more pure semiconductor materials.[26] This sounded complicated enough for people who could easily read and write English. Ken Reightler was concerned how things would go with a Russian on the crew.

All during these preparations there was an inside joke between Reightler and Dick Covey. Since both men were handling the details on welcoming the cosmonauts to Houston, one of them was likely to be assigned to the mission featuring the first Russian crew member. Who that would be became apparent when Covey was assigned to the first repair mission for the Hubble Space Telescope. Ironically, another crew member for that flight was Claude Nicollier, the European astronaut who needed an official escort to the simulators at the Johnson Space Center. Reightler was named pilot on the flight of STS-60, and the cosmonaut who would go along was Sergei Krikalev.[27]

The first Russian to serve aboard a NASA Space Shuttle had the distinction of launching to the *Mir* space station from one country and landing in another. Krikalev's mission began with a routine blastoff from Kazakhstan on May 18, 1991. His return was delayed by the disintegration of the Communist Soviet Union and its transformation to the Commonwealth of Independent States. The five-month extension to his flight kept him in orbit for 313 days. The Russia to which he returned would soon be teaming up with the United States, whose space program dealt such a stinging defeat with a successful Moon landing in 1969.[28]

Krikalev, his wife, Elena, and daughter, Olga, would be under Ken Reightler's wing as they got used to life in the United States. The

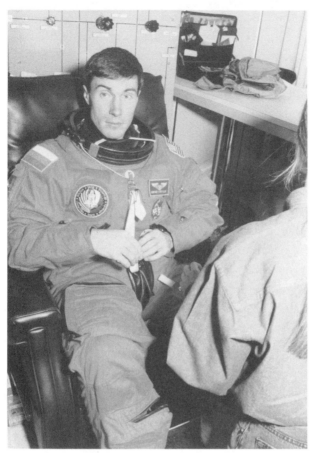

Cosmonaut Sergei Krikalev suits up for his mission as the first Russian crew member on a NASA Space Shuttle flight. The six-member crew deployed the dish-shaped Wake Shield Facility and conducted other experiments during the trip to Earth orbit. Courtesy of NASA.

veteran astronaut met his cosmonaut counterpart at the airport, wondering what to do first to introduce his Russian guest to life among former Cold War foes. The day after Krikalev arrived was election day, and Reightler figured that was a good place to start. Both men waited in line for Reightler to cast his vote. Krikalev's family arrived soon after, and that led to another big tradition later in the month when everyone sat down to Thanksgiving dinner.[29] Not everything went smoothly during the transition, however. The cosmonaut's first trip to the grocery store to stock his apartment stands out in Reightler's mind. "Being an engineer, I thought we could just break all this down to breakfast, lunch, and dinner," he says. What did the Krikalevs like for breakfast? Cereal, they responded. A trip to the cereal aisle at the local Kroger's, with the overwhelming selection of colorfully boxed brand names, took some getting used to.

Training for the upcoming joint space mission turned out to be easier than picking between *Wheaties* or *Cap'n Crunch*. The Shuttle program had evolved since Reightler's first trip to orbit in 1991. Back during his rookie flight, Reightler and his crewmates had a single laptop computer. It performed the simple task of providing a rotating map of Earth as the spacecraft circled the globe. On the planned mission of *Discovery* in 1994, there would be seventeen laptops crowding the crew cabin and the Spacehab laboratory module nestled in the Shuttle's cargo bay. Here, Sergei Krikalev was in his element, and mission planners intended to utilize that skill. NASA made him "Mister Laptop" for the duration of STS-60.[30]

Along with Reightler and Krikalev, the crew of *Discovery* included Commander Charles Bolden and astronauts Jan Davis, Ron Sega, and Franklin Chang-Diaz. Their résumés included previous international missions. Bolden was the copilot on a Shuttle flight in 1992 that included a crew member from Belgium. Davis flew aboard Space Shuttle *Endeavour*, which carried the Japanese Spacelab-J mission and crew member Mamoru Mohri, who was hailed as Japan's first professional space explorer, despite the launch of a Japanese journalist on a Russian Soyuz in 1990. The inclusion of a veteran Russian on the Shuttle was an eye-opener for the NASA "veterans" on the mission of *Discovery*.

Krikalev's first mission was in 1988 to the *Mir* space station. He later did back-to-back tours aboard the older orbiting outpost during the disintegration of the Soviet Union. Next came the much shorter trip aboard *Discovery*.[31] The cosmonaut may have been a rookie among the Shuttle astronauts, but only as Babe Ruth was a newcomer when he went from the Boston Red Sox to the New York Yankees. This wasn't unnoticed by Ken Reightler, who was in charge of getting Krikalev ready for this, his Shuttle trip. "I had just finished my first mission, which was five and a half days," he recalled. "I was going to train Sergei, and he had fifteen months in space. Who was training who?" A lot of the work included questions on if a certain emergency occurred on the Shuttle, how would a cosmonaut handle a similar situation on the *Mir* space station? Krikalev was also trained how to operate *Discovery*'s long robotic arm, making him a full-fledged member of the crew.[32]

Traditions

Prelaunch rituals play a role prior to blastoff, whether it's at the Kennedy Space Center or the Baikonur Cosmodrome in the former Soviet nation of Kazakhstan, and that applied to the countdown of Space Shuttle *Discovery* on STS-60. Back in 1961, when cosmonaut Yuri Gagarin was driven out to his spacecraft for the world's first manned orbital flight, he took time to urinate on the back tire of the bus.[33] Russian space fliers who followed in his footsteps christened their buses in similar fashion. Astronauts on Shuttle flights have their rituals as well. Frequently, they'll put a sticker of their individual mission crew patch somewhere, and the astronauts are instructed to "tap" the patch as they pass by for good luck. Launch controllers have a more culinary tradition following each liftoff. Beans and cornbread were first served following the first launch of the Space Shuttle in 1981. The legend began with NASA test director Norm Carlson, who brought a pot of beans for the support crew. Following subsequent launches, more and more staffers showed up, bowls in hand. Eventually, the demand required sixty gallons of beans be cooked up to satisfy everyone.[34] The day of the liftoff of STS-60, and the first Russian Shuttle crew member, prompted a brand new tradition in the room where the astronauts zipped up their spacesuits.

A Snort before Launching

The process of preparing for each liftoff includes a wake-up call for the astronauts at the Kennedy Space Center's dormlike crew quarters. The group gathers at a breakfast for a traditional photograph. A sheet cake adorned with their mission insignia is put on the table before them. The one for Reightler, Krikalev, Bolden, Sega, Davis, and Chang-Diaz was similarly decorated with the circular patch design.[35] The Space Shuttle was depicted in the center, the crew members' surnames were written around the border, and birdlike wings stretched wide on either side of the spacecraft. One wing was colored with the stars and stripes of the U.S. flag, and the other in the red, white, and blue bars of the flag of Russia. As tradition holds, the astronauts can look at the cake, but no one gets a slice. Usually, the crew eats light to avoid becoming nauseated in orbit.[36]

The next step is for the astronauts to suit up in their ninety-pound "dayglow" orange pressure suits. NASA support staff members gather in a room with big leather easy chairs to help each of the astronauts put on the bulky suits. The morning of *Discovery*'s launch, Ken Reightler sat opposite Sergei Krikalev in their matching chairs for suit-up. "He was just as cool as a cucumber," recalled Reightler of his Russian crewmate. The support team contributed to the light mood with an "icebreaker" of its own. The one-piece spacesuits are laid out flat on a table with the boots, gloves, and fishbowl helmets. Alongside the suits are all the supplies and tools that go into various pockets. "There are pens, pencils, chemical glow sticks, little viewing mirrors, and survival gear," says Reightler. "It's your last chance to know where everything goes." Each crew member's supplies are marked with a color-coded dot, with one color for each astronaut. Along with the typical items for the astronauts, the suit-up crew had included small souvenir bottles of vodka, each marked with the crew's colored dots. "We all sort of knew those weren't going along for the flight," joked Reightler. The astronauts later boarded the Space Shuttle with Krikalev seated directly behind Commander Bolden and Pilot Reightler on *Discovery*'s flight deck. The mission proceeded without incident.[37]

Ken Reightler left the NASA astronaut corps in 1995 to pursue a management position with Lockheed Martin. The aerospace company would go on to win the contract to build the agency's next-generation spacecraft, the gumdrop-shaped Orion capsule. Reightler wouldn't get to fly the new vehicle, but he would help with the design process.

The relationship he built with Sergei Krikalev would last beyond the conclusion of their joint mission aboard Space Shuttle *Discovery* in 1994. One tangible sign of that came on Halloween night in 2000. The cosmonaut had been named one of the three crew members who would work aboard the nucleus of the International Space Station. At that point, the outpost was comprised only of the drum-shaped *Unity* compartment, built by the United States, and the school bus–size *Zarya* compartment from Russia. Krikalev and his crewmates, Yuri Gidzenko and William Shepherd, blasted off aboard a Soyuz space capsule from the steppes of Kazakhstan. According to Russian tradition, the crew's families were not allowed to attend the liftoff.[38]

That prompted a telephone call from Moscow to Ken Reightler's office in Houston, Texas. Krikalev's wife, Elena, and his daughter, Olga, had attended both of his Shuttle launches from Florida. That included the 1994 *Discovery* mission with Reightler and a subsequent trip aboard *Endeavour* in 1998 that carried up the first part of the International Space Station. Krikalev had not blasted off aboard a Soyuz since 1991, and would miss not having his family on hand to see him off. "Sergei asked me to come to Baikonur to represent his family who couldn't come," says Reightler. "The next day I went to Moscow to share with Elena and Olga so they could experience it vicariously." After that flight, Krikalev would be commander on a later expedition aboard the orbiting outpost before retiring to a management position at Russia's rocket builder Energia. He and Reightler would stay in touch by e-mail and telephone.

Despites the successes of foreign participation on the Space Shuttle and the International Space Station, the debate drags on whether this kind of cooperation would work on a manned trip to Mars. Still, visionaries and mission designers are looking at how to send people to the red planet and what they might do there once they arrive.

getting there, living there

The challenge of sending astronauts on a mission to Mars includes unanswered questions for NASA. The dangers appear obvious. Although visiting the red planet will end the era of shorter, vacation-style spaceflights, scientists eager to visit Mars are making a list of sights they want to see that resembles the itinerary of someone on holiday. There's also another subject to be addressed.

Who will take the first step onto Mars?

Assuming the crew members who make the journey are in their thirties or forties, these future astronauts are in grade school now. They're about the same age former astronaut Scott Horowitz was when Neil Armstrong and Buzz Aldrin walked on the Moon during Apollo 11 in 1969. In fact, NASA was launching its lunar landings so frequently back then, the sixth grader from Philadelphia was worried his chance to go to Mars was slipping away. "Looking at where we were, we went from here to here toward the Moon, I thought we'd be on Mars in another five or ten years," Horowitz says. "It was the next obvious step, and then we stopped." That same brand of anxiety may be plaguing the next generation of astronauts waiting for their chance to fly to Mars now that NASA's interest in the planet has been renewed.

As for who will be the first person to walk on Mars, NASA won't know until the astronauts are selected and trained. But if the historic landing of Apollo 11 is any indication, the agency won't really know until someone's boot actually touches the surface. Even during the descent of the lunar module *Eagle* on July 20, 1969, no one was sure how it would turn out. That included astronaut Alan Shepard.[1]

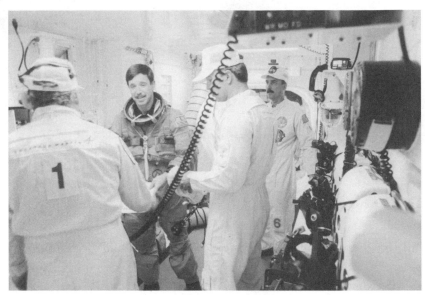

Astronaut Scott "Doc" Horowitz prepares to strap in aboard the Space Shuttle *Columbia* for the second mission to deploy the Italian tethered satellite. He would act as pilot of this mission and later command a servicing mission to the Hubble Space Telescope. Courtesy of NASA.

America's first man in space was chief of the astronaut office from 1963 to 1974, except when he was training for his own Moon landing aboard Apollo 14. Shepard was in mission control in Houston the evening of Neil Armstrong and Buzz Aldrin's final approach to the lunar region known as the Sea of Tranquility. Technicians sat behind banks of computer screens, many with the same thought on their minds: will the lunar module work?[2]

After blasting off aboard America's first Mercury capsule in 1961, Shepard was temporarily removed from active flight status by an inner-ear problem. Surgery later cleared him to return to active duty. In the meantime, he was put in charge of training the other astronaut crews as NASA got closer and closer to the first Moon landing. He was familiar with Apollo 11 crew members Buzz Aldrin and Michael Collins, but especially the mission's commander, Neil Armstrong. "I really remember Neil from a long time ago because I flew at Edwards [Air Force Base] on several navy projects when he was out there," Shepard recalled in 1994. "He was quiet, reserved, a good 'stick and rudder' man. He was a hero in the mold of a Lindbergh. He [Lindbergh] was a pretty quiet individual." Shepard remembered how, as the moments ticked closer to the planned Apollo 11 lunar landing, no one in particular was singled out from the start to be the first man on the Moon.

The crew of Apollo 9, astronauts Jim McDivitt, Russell Schweickart, and David Scott, blasted off on March 3, 1969, to test the lunar module *Spider* in Earth orbit. Astronauts John Young, Gene Cernan, and Tom Stafford followed that flight up with Apollo 10 on May 12, 1969. That crew's lunar module *Snoopy* traveled to within 50,000 feet of the surface of the Moon before firing its engines in an emergency abort to confirm the safety option would work.[3] Shepard asserts that only because these early flights worked fine did Apollo 11 get the "go ahead" to land during its mission. "The crews were selected primarily on who was best qualified at the right time, not for any specific flight in mind for the landing," says Shepard. "Apollo 10 indicated things were going

Astronauts Neil Armstrong, Edwin "Buzz" Aldrin, and Michael Collins rest inside a mobile quarantine facility aboard the aircraft carrier USS *Hornet*. They were kept in medical isolation following their historic mission to make the first manned landing on the Moon. Courtesy of NASA.

pretty well, but Apollo 11 might not have made it, and then Apollo 12 might have made it." Armstrong's landing is etched firmly into the history books, but during the final seconds, computer error messages from the *Eagle* and the dwindling fuel supply during the approach had mission control concerned. "When you have any kind of problem, whether you're in the control center or in the spacecraft," Shepard recalled of that night, "your reaction isn't one of 'oh my god we have a failure.' You think about how we work around it, what switch do we throw, what alternate scheme do we bring into play so we can continue on a reasonable basis." The first Moon landing played out as schoolkids remember it today.

That left Shepard to deal with stories of how NASA supposedly wanted Armstrong to be the first to walk on the Moon because he was a civilian aviator. Aldrin, by contrast, was a decorated air force fighter pilot who shot down two Russian MIG jets during the Korean War.[4] "Fairy stories, fairy stories," says Shepard with a wry smile. "The reason was that the door of the lunar module was hinged on the right-hand side. It was physically impossible for Aldrin in his heavy spacesuit to get around Armstrong and out the door." The alterative appeared to be either spend millions of dollars to change the door, or let Armstrong go first.

The uncertainty of the success of the Apollo 11 Moon landing also led to a precaution that may be used during NASA's possible future touchdown on Mars. William Safire, an aide to Nixon adviser H. R. Haldeman, was asked to prepare a speech for the president in case the astronauts crashed on the Moon, or their lunar module failed to lift off, leaving them stranded with no chance of rescue. "Fate has ordained that the men who went to the Moon to explore in peace will rest there in peace," the unused statement read. "These brave men, Neil Armstrong and Edwin Aldrin, know there is no hope for their recovery. But they also know that there is hope for mankind for their sacrifice. These two men are laying down their lives in mankind's most noble goal, the search for truth and understanding."[5] Proponents of a mission to Mars admit that astronauts may be lost along the way. NASA has even begun studies on the sensitive issue of under what circumstances would an injured crew member be allowed to die if it threatens the mission or the rest of the people on the spacecraft. Robert Zubrin is among those looking at this scenario pragmatically. He's a former engineer with Lockheed Martin and founder of the space advocacy group called the Mars Society. "No great project ever happened without risk," Zubrin asserts. "Men died building the Brooklyn Bridge, and men certainly died on the beaches of Normandy." Designers and dreamers of NASA's first manned venture to Mars are thinking of ways to give the astronauts their best chance at survival and what they will do once they are there.

Assuming Congress is willing to pay the hundreds of billions of dollars the mission might cost, the spacecraft would likely include some new type of crew capsule. The astronauts would use the new vehicle to fly to Earth orbit, and back to the ground at the end of the mission. Bigger booster rockets will carry up some kind of habitation module for the trip to and from Mars, a cluster of engines to propel the vehicle during the round-trip, and a landing craft for the perilous voyage to the surface of the planet.[6] There are many possible variations that could be snapped together in Earth orbit for the voyage. Europe might build the crew compartment. Russia and Japan might supply booster rockets for the mission.

The challenges associated with a trip to the red planet were on the front burner for President Obama's Augustine Commission, which mapped out options for the future course at NASA. The panel's primary concern, beyond the monetary cost, was the danger that cosmic radiation would pose to the astronauts. Earth's atmosphere helps to screen out the stream of deadly charged particles from the Sun. That natural shield doesn't exist in the void between Earth and Mars, which could leave the crew of a spacecraft vulnerable to exposure. The Martian atmosphere is also very thin, so the cushion of air that reflects much of the ultraviolet light striking Earth isn't present on the red planet, creating the ultimate problem of getting people to and from Mars as quickly and safely as possible. Mission planners also want at least some redundancy in the vehicles to ensure a reliable ride there and back.[7] In broad-brush terms, there are two basic plans to achieve the goal.

Flying "Business Class" or "Economy"

NASA visionary Robert Zubrin, the Mars Society, and the National Space Society have all pondered the questions of a Mars mission. The result is two competing strategies called NASA's "Reference Mission" and "Mars Direct." Both involve launches with and without crews on rockets designed specifically to carry people or cargo, either directly to the surface of Mars or to one of its two moons.

NASA's early strategy involves four separate launches for each expedition. The mission would require some kind of heavy lift launch vehicle, akin to the Ares-5 cargo rocket envisioned by the space agency following the 2003 *Columbia* disaster. Each of those muscular boosters would carry a smaller rocket designed to propel people or materials from Earth to Mars. These vehicles would use nuclear-powered engines descended from the early tests in NASA's Nuclear Engine for Rocket Vehicle Application (NERVA) project in the 1960s. The first launch would deliver a fueled spacecraft known as the Earth Ascent Vehicle, or EAV, to orbit around Mars. If nothing else, the success of the first flight would give mission managers and the astronauts the psychological reassurance that if the mission were to go sour, there would be some way for the crew to get back home.[8]

The second launch would send a departure ship, called the Mars Ascent Vehicle, or MAV, to the surface of the planet, along with a habitation module and a nuclear generator. The third blastoff would propel a second crew module as a safe haven to the planet along with a backup generator and survival supplies.

The "Reference Mission" is the slow and steady path to sending humans to Mars. Only after the unmanned base is up and functioning on the planet during one launch opportunity would the astronauts be sent on a six-month-long voyage to Mars during the next close-proximity pass with Earth two years later. Additional cargo launches would also be made to support the follow-up crew set to go on a later mission.[9] NASA's "hopscotch" idea would ensure that the astronauts would have the tools they need for survival.

Robert Zubrin and the Mars Society, however, favor a more "quick and dirty" approach called "Mars Direct." That plan trims the number of launches to three and tightens the schedule. With "Mars Direct," the first large cargo rocket would send the EAV directly to the surface of the planet, along with a nuclear generator and an automated chemical factory to create oxygen from the carbon dioxide in the atmosphere for life support and fuel. The second and third launches would occur two years later, during the next Earth/Mars close encounter. One vehicle would carry the astronauts, and the other would include a redundant return craft. This would be the first in a series of "leap-frogging" follow-up missions, each about 250 miles from the previous expedition, to explore a wider range of Mars.[10] With the basics of the trip covered by these two strategies, other visionaries have the latitude to be a bit more fanciful in terms of what a voyage to Mars should include.

Artificial Gravity, Yes or No?

While NASA is already thinking about the technical and psychological demands of a trip to and from the red planet, retired astronaut Scott Horowitz is worried about crew survival once they land.

The manned entry vehicle, or Martian lander, is expected to use a combination of heat shielding and parachutes to slow itself down through entry into the thin atmosphere of the planet. Then retro rockets could be utilized to cushion the vehicle as it settles down on the surface.[11] The concern is what if something goes wrong and the crew lands off target, miles from the safe haven of its habitat module? In Horowitz's mind, that's where sending the crew without some kind of artificial gravity could be fatal.

Astronauts on long, weightless space missions lose bone and muscle mass due to the lack of gravity. Their weakened condition is similar to the lack of bone density that the elderly often face. "That brings up the whole debate of whether you want the crew to be de-conditioned when they land," says Horowitz. "What if they had to do something right after landing because there was a problem? Now the crew may be in no condition to help itself." Astronauts left weak due to a long trip under zero gravity could also force NASA to make design compromises in its landing craft. Crew members aboard the International Space Station return to Earth aboard the Space Shuttle in flight seats that allow them to make the trip lying down to avoid injury or fainting as gravity returns. De-conditioned astronauts on their way down to the surface of Mars could be required to complete the landing flat on their backs instead upright and looking out the windows like the Apollo Moon crews.[12]

The solution could mean finding a method of creating gravity. Rotating the main spacecraft during the voyage to Mars to reduce the feeling of weightlessness among the crew members is one concept being considered. Proponents of a spacecraft design including modules connected in a long line say the crew could live in the outermost compartment as the whole complex cartwheels its way from Earth.[13]

The unanswered question is how much gravity would be enough? One theory is that replicating the force on the Moon, which is one-sixth as strong as Earth's, might be enough to maintain the astronauts' health. Robert Zubrin compares it to the situation the military faced during World War II, when bomber pilots flying above 20,000 feet began blacking out due to the lack of oxygen. Doctors suggested medications to help the crews adapt to the thin air, while others championed the use of oxygen masks to allow the pilots to breathe. The masks won out. "We've all been living under gravity," says Zubrin, comparing the military oxygen problem to NASA's gravity question. "Trying to change human physiology is hugely complex compared to rotating the spacecraft." The debates will likely drag on as the agency irons out the details in its plans to send people to Mars.

A Day in the Life on the Red Planet

Just as the spacecraft that carries NASA's first Mars crew to the planet is subject to revision, so is the base that will sustain them on the surface. The space agency may opt for agriculture modules to grow crops and regenerate air and water. Inflatable crew modules could be used along with traditional metallic compartments to shelter the astronauts from the toxic atmosphere and the constant bombardment of radiation from the Sun.[14] Whatever structure is selected, proponents are already thinking ahead to what the astronauts' average workday will be and assembling a laundry list of spots on Mars that should be explored.

Reveille at 7 a.m.

A proposed workday, as envisioned by the Mars Society, would begin with a wake-up call for the astronauts at 7 a.m. Mars time. The planet rotates differently compared to Earth, so each day lasts twenty-four hours and forty minutes compared to a day on Earth. One solution is programming the wristwatches and clocks used at the Martian base with longer "seconds" to compensate, so the astronauts wake up to a sunrise on Mars each morning.[15]

The day might begin with a chance to clean up, followed by breakfast and the opportunity for the crew to chat before going to work. That could occur around 8 a.m. At 9 a.m., the team might gather in the main wardroom for a daily briefing with the mission commander. At that point the astronauts could be divided into two teams. One group would remain inside the habitat module to study rock and soil samples gathered during previous expeditions outside on the surface of Mars. Other members might spend the bulk of the day repairing equipment or sending reports back to scientists on Earth.[16]

The second group would head to the base airlock after the morning briefing and spend about an hour suiting up for a "Mars walk." This team of astronauts might drive around in a single pressurized rover with the crew sealed inside the cockpit to reach their destination, or perhaps smaller individual vehicles resembling dune buggies might be used for each member of the team. One advantage to the smaller cars is if one breaks down, astronauts could share the ride back to safety. Also, if a smaller Mars car gets stuck in a deep sand dune, the astronauts might be able simply to lift it out under the lighter force of gravity on the planet.

If the Apollo lunar landings are any indication, one thing future Mars explorers would have to get used to would be the complete silence during their sojourns from their home base. With their heads encased in airtight helmets, the early lunar crews couldn't hear anything like the sound of their feet crunching on the rocks in their path. This also left the astronauts uncertain that the precious photographs they were taking with their large Hasselblad cameras were actually being recorded. Unlike on Earth, there wasn't the reassuring "click" that was heard as each picture was taken. The Moon-walkers needed to keep an eye on the exposure counter to make sure the film had advanced for the next shot.

After spending the morning and much of the afternoon exploring their target territory, the expedition would head back to the base, unload their cargo of rocks and soil samples, and climb out of their bulky suits.

Next would be a de-briefing with the base commander, followed by dinner and some time off. The crew might read or sit down for a movie before "lights out" and bedtime around 11 p.m. This scenario might give the impression of a carefree workload, with astronauts bounding around under the lower Martian gravity, but the mission will include its forms of drudgery.[17]

When the International Space Station was limited to two or three crew members, a lot of the day was spent in basic maintenance on the orbiting complex. Likewise on a Mars base, routine housekeeping will be combined with improvised repairs on equipment to keep the crew alive and well during their stay on the planet. The astronauts will also have to cope with the environment on Mars and the dust that will contaminate the suits and the gear used outside. Still, scientists are already assembling an agenda of what they hope will be spectacular sites to visit on the red planet.[18]

The Wish List

John Mustard's office at Brown University in Rhode Island is unmistakable. The professor of planetary geology worked with NASA on the *Mars Reconnaissance Orbiter* spacecraft that circled the planet to gather data on familiar Martian landmarks, but also uncover new ones like buried glaciers. Mustard also helped choose possible landing sites for NASA's planned 2009 science rover, which was designed to visit more challenging terrain than its predecessors, the rolling robots *Spirit* and *Opportunity*. His office at the university is cluttered with posters and photos of Mars. Toy versions of the Mars rovers by Mattel sit on his desk, along with the "John Mustard" action figure created as a prank by some of his graduate students. "They took a Steve McQueen doll, and gave him a little shirt like one I wear a lot," he says. "And he's got a little beer can. I'm Canadian, so I must have beer." And clearly, Mars is a priority as well. Mustard has seen the red planet a lot through the eyes of NASA's *Mars Reconnaissance Orbiter*, and he has a list of places he'd like to visit firsthand, standing in a spacesuit on the surface of the planet. He calls it his "vacation list."

The Jewelry of Mars

When ordinary people look at rocks and mountains, they see rocks and mountains. Geologists see the passage of time. Scientists think Mars was once a warmer and wetter place billions of years ago, and may have been hospitable to simple forms of life. Now it's not. Mustard believes clues to how Mars changed are locked in the rocks, waiting to be studied. "I'd love to go to one of those early areas when Mars had a more clement environment that included the ingredients that could have supported life," he says.

The region known as Nili Fossae, for example, looks like a giant used his fingers to paw at the planet and leave a series of grooves in the surface. The longest fracture is about 300 miles long and 20 miles wide. This is also one of three locations where eruptions of methane gas created excitement in the scientific community starting in 2003. Methane can be a by-product of living things, which prompted speculation on bacteria subtly going about its business under the surface of the planet.[19] Mustard says just studying the terrain at Nili Fossae can reveal a lot about Mars. The gashes in the crust were likely caused by flowing water and might reveal layers of rock showing how Mars evolved. "I'd like to stick my hands into the dirt, and then walk from there through the time when all that changed," Mustard says, "when we thought Mars was a warmer and wet place to one that's really dry and really cold. I want to walk across that." Mustard believes one difference that will be seen by exploring the planet on foot is the variety of colors, especially in spots where windstorms wear away the yellowish surface soil. A little digging, he asserts, might uncover hues of blue, yellow, green, and deep red. "Not as dramatic as the painted desert," says Mustard, "but once you get past the surface soil, it will be wildly colorful." And then there are the semiprecious stones likely strewn across the surface of Nili Fossae. That region is believed to hold huge deposits of a green gemstone called olivine. The jewelry-grade version of olivine is called peridot, and Mustard contends there are vast fields of it at Nili

Fossae, like piles of lime-colored rock candy. "I think you could scoop up handfuls of this stuff," he says. "It would be a large and sparkly place." When astronauts and scientists get tired of exploring here, they could choose places more hazardous than the areas traversed by the *Spirit* and *Opportunity* rovers.

Scientists complain that, out of necessity, NASA had to choose safe landing spots for the rolling robots. The result was flat, and mostly boring, regions of the planet. Crew members on the surface of Mars could take more chances, like visiting Jezero Crater. This formation is believed to be an ancient lakebed with a river flowing into it, which formed a delta. People on Earth would see something similar if the Gulf of Mexico were suddenly drained of water, leaving only the soil deposits dumped at the mouth of the Mississippi River over millions of years.[20] "When we see these Mars deltas from orbit, there are fairly steep cliff faces with bands of sedimentary rock," says Mustard. "For geologists, that's pay dirt because that's environmental change and you can see it." Deltas on Earth traditionally include large deposits of clay, which can contain evidence of past life. Scientists think if astronauts are going to find definitive clues to ancient life on Mars, places like Jezero Crater could be a good place to look. "You're not going to find dinosaur bones sticking out," contends Mustard. "But the by-products of microbial life could be there, and that would be awesome." These by-products might not be fossils, but rather organic carbon left by these tiny Martians when they died off. Just as ancient life on Earth leaves dark stripes in layers of rock, life on Mars might be convincingly detected in the same way.[21]

The list of favorite exploring spots for astronauts on Mars doesn't include every landmark visible from orbit. Take Olympus Mons, for example. It's the largest known volcano in the solar system, as big as the state of Arizona. It's also three times as tall as Mount Everest on Earth. The European Space Agency's *Mars Express* spacecraft took spectacular views of Olympus Mons in 2004. Photographs of its northern flank included dramatic landslides, which may be the result of glacial activity.[22] Still, even enthusiastic supporters of a manned Mars base doubt that Olympus Mons would be worth visiting up close. It's so big, they contend, that the crew members walking on the surface would "lose the forest for the trees." Astronauts might only see the ground sloping gently upward for miles on end with nothing particularly significant about it.

Valles Marineris, on the other hand, seems to make scientists uniformly excited.

This huge canyon along the Martian equator is long enough to stretch more than one-quarter of the way around the planet. On Earth, its length is the distance from New York to Los Angeles. It's also five miles deep, making the Grand Canyon seem more like a ditch. Some mission designers suggest sending expeditionary teams up to the rim of the yawning chasm, only to send small robots down for remote exploration.[23] John Mustard wants to do something different at the very beginning of a given trip to Mars. Instead of going to the edge and looking down, he suggests landing the astronauts down in Valles Marineris so the crew can get out and look up. "You'd probably get a sore neck because you're looking up all the time," Mustard says. "But that's where the cool stuff would be." This kind of landing could be especially dramatic, with the astronauts seeing the sheer cliffs of the canyon all around them as they come in on final approach. Their spacecraft would likely appear to land "in" the planet, rather than "on" it.

Assuming NASA attempts a landing on Mars at all, the dreaming has to end and the process of picking a specific landing site has to begin. That's where mission planners and the scientific community could find themselves at odds.

Tough Choices

NASA's proposed plans to send astronauts back to the Moon are focusing on places like Shackleton Crater at the lunar South Pole. Mission managers favor the relatively moderate temperatures there with constant light from the Sun to help generate solar power. The crew might also find ice in the shadows of craters that could be used for drinking or the generation of oxygen for life support or rocket propellent.[24] Members of the scientific community are already gearing up for a fight in case the first Mars base is chosen solely on the basis of habitability, with little regard to what the astronauts can reach and study. Robert Zubrin with the Mars Society contends that if water is the concern, NASA doesn't have to limit itself to the North or South poles where ice appears abundant, but areas of scientific interest are sparse. He points to the Viking missions in the mid-1970s that visited Utopia Planitia and Chryse Planitia in Mars's northern hemisphere, which appears to have frost. "Viking found soil with 3 percent water by weight, and some up to 60 percent water by weight," Zubrin asserts. The question of choosing a landing spot based on practicality over scientific merit has John Mustard worried as well. A polar landing would likely put his "vacation list" including Valles Marineris and Nili Fossae out of reach of the astronauts. However, he contends, there are points of interest at the poles if that's where NASA chooses to go. A different canyon, called Chasma Boreale, cuts into the ice of the north polar cap with sheer cliffs almost a mile high. "If you were standing at the bottom of that, looking up, you'd see all climatic variations laid out in those layers, like the ice sheet in Antarctica," says Mustard. "It would be otherworldly." But it would also be an uncomfortable compromise. If NASA wants to study the possibility of life on Mars, he contends, the agency will have to get people to the southern highlands.

At this point, arguing over landing spots would be an aerospace version of "counting your chickens before they hatch." NASA still hasn't returned to the Moon, and there are critics who say sending people to Mars would be dangerous, expensive, and of questionable scientific value.

why go at all?

The debate over the human risk and the breathtaking cost of sending astronauts to Mars began long before the end of the Space Shuttle program. It even predates the birth of President George W. Bush's vision of traveling back to the Moon and beyond. It goes back even farther than the first launch of Shuttle *Columbia* in 1981.

In 1970, when members of Congress were arguing over the fading luster of Project Apollo and NASA's manned Moon missions, a flight to Mars was on the table. There was also a proposal to build the Shuttle or a space station, or to send up to six of the future unmanned Viking landers to the red planet. Seething through all the speeches was Senator Walter Mondale of Minnesota. He, along with fellow lawmakers like William Proxmire and James Fulton, was among the staunch opponents of spending billions of dollars on sending astronauts anywhere. During a debate over spending $110 million of the taxpayers' money to fund the design of the Shuttle and space station, Mondale tried numerous times to strike the money from the budget. "I believe it would be unconscionable to embark on a project of such staggering cost when many of our citizens are malnourished, when our rivers and lakes are polluted, and when our cities and rural areas are dying," he said. "What are our values? What do we think is more important?"[1] The space station went unfunded, the troubled Space Shuttle program barely survived, and no manned Mars flight would be attempted.[2]

Décades later, the idea of launching manned vehicles to Mars is being met with similar opposition. It basically breaks down into three groups: those who hold fast to the belief that going to Mars is too expensive, that it's too technically challenging, or that robotic missions would do a better job of studying the planet.

A Tale of Three Critics

Nobel laureate Steven Weinberg's interest in science was the result of a "hand-me-down." He grew up in New York City, and one of his cousins had a chemistry set he didn't want, so Steven got it. "It had a lovely wooden box, and pretty chemical glassware," Weinberg recalled. "I got fascinated with it in a childish way. I liked to make bangs and stinks." That included concocting small amounts of gunpowder, among other things. He also found that science didn't give up its secrets easily, and you had to work to uncover subtle truths about how the world worked. Instead of discouraging him, it sent Weinberg on a path toward his doctorate in physics from Princeton.

In 1967, while lecturing at Massachusetts Institute of Technology, Weinberg laid the foundation of his work on weak electromagnetic bonds on elementary particles that eventually earned him the Nobel Prize in 1979. During this time, Weinberg also first encountered NASA. The space race against the Soviet Union was on, and the general public was more aware of science in general. "Even my relatives were looking at me with more interest," Weinberg recalled. "I was in science, and science all of a sudden was of importance." The Apollo astronauts did follow through on this surge of mainstream enthusiasm. They brought back lunar rocks and set up reflectors on the surface, so scientists on Earth could bounce laser beams off them to chart the course of the Moon. For Weinberg, that was useful. "You could follow the orbit of the Moon to a fantastic accuracy, within inches," he says. "You could test Einstein's theory of relativity when the motion of the Moon matches that theory." But that was the extent of the scientific payoff from sending people into space. In Weinberg's view, NASA's limited technological reach in the 1960s made it necessary to send humans to achieve these objectives. Now, robotic flights could do the same thing more effectively.

Two years after Weinberg collected his Nobel in Stockholm, NASA launched the first Space Shuttle. The professor's opinion on the value of manned spaceflight continued to deteriorate. "I'm not opposed to sending people to the Moon or Mars for its own sake," he says. "I get a natural excitement over it like most people. But it's not a good way to do science." The space program did make its mark, in Weinberg's view, with unmanned spacecraft. One example is the Cosmic Background Explorer, or COBE, which was launched in 1989 aboard an Atlas rocket. The squat boxy satellite, with its twin solar panels, was designed to look for the leftover ripples of energy from the so-called Big Bang, which astronomers believe created the universe. After making contributions in the area of physics, Weinberg's interests branched out into cosmology, so COBE's findings were of particular interest. "It provided a wealth of remarkably accurate data," he says. "It found the universe is thirteen billion years old with a certainty of one hundred million years. That's less than a percent." COBE's results also helped two of its primary mission scientists win a Nobel Prize of their own.[3]

Still, the lack of scientific "bang for the buck" from sending astronauts to space, compared to robotic missions, continued to be apparent to Weinberg. More bothersome was the apparent trend in Congress of funneling money toward manned space trips at the expense of projects of greater scientific promise. This was no clearer than during the aftermath of the *Columbia* disaster in 2003. President George W. Bush called on NASA to pursue a new vision of sending astronauts to the "Moon, Mars, and beyond." Once everyone got past the catch phrase, scientists like Weinberg noticed an immediate chill on funding new unmanned exploration projects so that the Orion capsule, which was intended to replace the outmoded Shuttle, could take precedence. "NASA announced they were canceling science programs," says Weinberg, "because it doesn't advance the president's vision of going to the Moon or Mars." Recommendations to the agency during this time included axing unmanned spacecraft to study asteroids, orbit three moons of the planet Jupiter, and scan Earth's atmosphere.[4]

Bush's plan to retool NASA appeared to give the space agency a new mission. But Weinberg believes the idea of exploring Mars shouldn't be done with astronauts, but with robotic probes. "For the cost of sending one manned mission to Mars, you could launch hundreds of robotic probes to hundreds of locations on Mars," he says. "You don't learn much by exploring just one tiny patch, and it would be better to [go to] a large variety of different locations."

Also, human beings can get in the way of science. The presence of astronauts, scientists complain, can create heat or vibration that could interfere with the readings of sensitive astronomical equipment.[5]

Is the Trip Technically Possible?

When it comes to problems related to NASA, policy analyst Gregg Easterbrook has been saying "I told you so" for thirty years. In April 1980, he wrote one of his first articles that were critical of NASA. The piece for the *Washington Monthly* compared NASA'S new Space Shuttle to the controversial post–World War II–era flying boat known as the *Spruce Goose*. Billionaire and aviator Howard Hughes built the $25 million aircraft as a troop carrier for the U.S. military. Seven hundred soldiers were meant to ride aboard the 200-ton vehicle, assuming it could really fly. Hughes took the controls of the prototype airplane on a test flight from Long Beach harbor. It traveled one mile, at an altitude of seventy feet, before plopping back into the water to the sound of creaking and snapping timbers.[6]

Easterbrook wondered if NASA's fortunes would be any better during the era of the Space Shuttle. The billion-dollar vehicle was a full year away from its maiden flight with veteran astronaut John Young and rookie pilot Robert Crippen on board. Easterbrook's feature was printed the year after the producers of the James Bond movie series endorsed NASA's space plane by featuring it in the film *Moonraker*. Industrialist and Bond villain Hugo Drax, played by actor Michael Lonsdale, stages the improbable theft of a Space Shuttle. A pilot is hired to hijack the vehicle by rocketing it off the back of its Boeing 747 carrier jet. The lack of an external fuel tank or solid rocket boosters needed to actually propel the spacecraft didn't deter the moviemakers, who simply deleted the hardware for the sake of storytelling. Later, agent 007, portrayed by veteran actor Roger Moore, would board one of a fleet of Shuttles filled with Marines armed with lasers and jetpacks to foil Drax's evil plan to kill Earth's population with deadly satellites. Whatever the reaction might be to the movie's plot, it depicted the Space Shuttle as a viable descendant of NASA's Apollo program.[7]

Easterbrook's article did not.

Along with pointing out the cost overruns and lack of jobs for the actual Space Shuttle to do in orbit, like rescuing and retrieving damaged satellites, he also joined the ranks of critics of the use of solid rocket boosters to begin the trip to orbit. No one had ever used solid fuel to launch human beings, because the boosters couldn't be turned off in an emergency once they ignited. The rockets would simply burn until the propellent is exhausted. "Solid rockets can fail in two ways," Easterbrook wrote. "They can explode. Or they can shut down spontaneously. If a booster shuts down, there will be 2.5 million pounds of thrust on one side battling zero pounds on the other. Even a split second of this imbalance will send the ship twisting into oblivion, overriding any application of pilot skill." Six years later, a failed rubber gasket on a booster attached to Space Shuttle *Challenger* would cause NASA's deadliest disaster at that time.[8] Easterbrook remains amazed that the space agency was given the "go ahead" to build the Shuttle in the first place. "NASA was the best example of the 'can-do' spirit," he says. "In the 1960s, their track record was phenomenal. People still had the impression of NASA as a *do no wrong* agency. After two Shuttle disasters, we don't have that anymore." During ten years with the Brookings Institution in Washington, D.C., Easterbrook had the opportunity to watch NASA over the long haul, and he believes it's too technically challenging for the agency to send people to Mars.

The proposed trip would be far beyond what the agency had previously attempted with Apollo, the Shuttle, or the International Space Station. One danger that astronauts have previously not faced is excessive radiation from the Sun or from cosmic rays. That's one of the top concerns on Easterbrook's list. Crews in low Earth orbit are largely shielded from harmful radiation exposure by the planet's own magnetic field. Even on brief trips to the Moon, astronauts on Apollo weren't out in space long enough for solar flares streaming from the Sun to threaten them. However, during the long trip to Mars and the months spent on the planet's surface, the perceived danger would be much greater. The spacecraft would spend long periods in transit between Earth and Mars, where no natural protection would be available. The thin Martian atmosphere and the lack of a magnetic field around the planet would leave the crew exposed as well. Solutions to the radiation problem, in Easterbrook's view, would make such a voyage nearly impossible. One obvious answer would be to provide shielding for the crew, and the most obvious material is lead. "When you do a rough calculation," he says of the weight of the spacecraft, "you just throw up your hands." It took a 50-ton vehicle to send people to the Moon during Apollo on trips lasting no more than twelve days. By Easterbrook's reckoning, a

Mars craft would have to weigh up to 10,000 tons to carry enough fuel and provisions for the mission. "That's the weight of the guided missile cruiser *Ticonderoga*," he asserts. "We are not sending the *Ticonderoga* on a mission to Mars anytime soon."

Then there's issue of propelling the vehicle, where Easterbrook believes NASA has no good engine solution to keep a manned mission to Mars feasible. He proposes shelving the idea of a trip to Mars in favor of reordering the space agency's priorities. He suggests focusing on propulsion research until a breakthrough can be found that lowers the cost and the weight of a protracted Mars flight. Along with that, Easterbrook suggests a trimming of the number of people currently occupying the astronaut corps, and renewed attention to what he believes could be NASA's biggest contribution to humankind—devising a way to protect Earth from asteroid strikes. "It would cost only a few billion dollars," he believes. "And if they save a major city from a major hit, that would be the greatest achievement in human history." In summation, Easterbrook believes NASA has been adrift since 1972 and the conclusion of the lunar landings. His criticism isn't limited to managers at the space program but is also directed at lawmakers in Congress who govern NASA's actions and its budget, which amounts to 0.5 percent of federal spending. "If they [Congress] can't get right one of the smallest expenditures, how can they handle the bigger stuff?" he asks.[9] It's the cost of manned spaceflight that comes up most regularly among NASA's biggest detractors. Even people who like the idea of it are worried about the price tag.

No Bucks, No Buck Rogers

The day before Halloween in 2008, NASA had more on its mind than trick or treat. The agency's authorization bill was up for debate before the U.S. House of Representatives. The document sets out NASA's objectives to be funded through the appropriations process at a later date. To keep the Halloween analogy going, it's like bobbing for apples in the form of dollars from Washington while the space program's critics line up to swat the agency like a piñata. This procedure sets NASA's agenda, and the evolving power structure in Washington has a direct impact on how NASA goes about its business.[10] The Eisenhower administration, in the late 1950s, wanted NASA to be a civilian agency, so its funding could be carefully scrutinized. The Kennedy White House set the goal of sending astronauts to the Moon. Following JFK's assassination, Lyndon Johnson tapped the brakes on Apollo's spending, as did Richard Nixon. This gave birth to the Space Shuttle Program. Gerald Ford oversaw the Apollo-Soyuz Test Project, which concluded the manned lunar program. Jimmy Carter presided over the development of the Shuttle, which first launched during Ronald Reagan's watch. The Reagan years included the proposal of space station *Freedom*, which later evolved into the International Space Station when Bill Clinton was elected commander in chief. The process of change for NASA would be no different in 2008, but the national landscape had changed drastically.[11]

The St. Patrick's Day Massacre

Seven months earlier, when Americans were celebrating another holiday, the feast day of St. Patrick, there were sobering headlines to go along with the green beer and parades. New York investment banker Bear Stearns had imploded during the nation's financial downturn, leaving 14,000 workers without jobs and another reason for the country's economic confidence to be shaken.[12] To longtime critics of NASA, it was another reason to renew their view that sending people into space was a hugely expensive enterprise and that tax dollars should be spent on something else.

Leading the charge against plans to send astronauts on a future mission to Mars was Massachusetts congressman Barney Frank. It was a long-held position. In December 2005, the Democratic lawmaker declared the notion of sending astronauts to Mars "flying pork." In 2006, Frank filed an amendment to completely prohibit the funding of a manned flight to Mars. "It's a complete and total waste of money," he told his colleagues on the hill. "The manned shot to Mars is pure boondoggle." The House passed NASA's funding despite the complaints, but Frank stuck to his guns as the economy began to weaken through 2008.[13]

When NASA's authorization bill came up for discussion in October 2008, supporters of the nation's space effort knew Barney Frank would be going after them. "I didn't make a secret of it," he says. "I told people when it came up, I was going to raise the issue." And he did. As discussions dragged on, Frank took the microphone to again speak out against sending people to Mars. "Space exploration is very important and has great scientific and practical results," he says. "But sending humans to Mars will cost hundreds of millions of dollars for very little scientific worth."[14] Frank insists his stance isn't against NASA, only against the expensive idea of putting people on Mars. Part of the federal economic stimulus package, created by Congress and the Obama administration to help the economy, included dollars for NASA in areas like aeronautical research. Supporters of a manned Mars flight say investing in such a mission would help the goal of reviving the economy. They point to the days of Apollo, when thousands of people where inspired to study engineering in college and to take jobs that led to putting men on the Moon.[15] Congressman Frank contends there are more cost-effective ways to get people to study science than going to Mars. "When people say that," Frank responds, "they're looking for ways to justify something they're already interested in on other grounds." Also, the incremental nature of the process of sending astronauts to the Moon and then to the red planet is a concern. If lawmakers don't have to worry about the costs up front, he contends, it's easy to support the idea and pass the fiscal responsibility along to future lawmakers.

Retired Republican congressman David Weldon had heard these arguments before. From 1995 to 2008, he represented District 15, which included NASA's Kennedy Space Center. As such, he was considered the "space guy" in the powerful House Appropriations Committee, which helped to dole out federal dollars to programs, including NASA. After announcing his departure from Congress to resume his medical practice, the November 2008 election would select his successor.[16] During the October debate over NASA's authorization bill, it was still Weldon's job to fend off opposition like Barney Frank's. He described the process as working lawmakers who were either "movable or immovable." Moveable members of Congress had no political stake in NASA's future, since there were no jobs hinging on space centers or aerospace contractors in their districts. Still, they liked the idea of spaceflight anyway. "They were the reasonable ones who wanted to hear the facts, and I spent a lot of time working with them," says Weldon. On the other hand, the Florida congressman found that immovable lawmakers disliked funding NASA's more expensive ventures, despite the favorable arguments from supporters. "They [the immovables] tended to be from urban districts," Weldon recalled. "They preferred spending money on education or other social programs, so I tended not to spend a lot of time with them." Reapportionment, which is the process of recarving congressional districts, eventually created a second seat representing the region around the Kennedy Space Center. That would give Weldon, and his successor in the House, a partner to lobby in favor of the residents surrounding NASA's Florida launch site. That was still far fewer than the dozen or so districts around Houston, Texas, which is home to the Johnson Space Center where the astronauts live and train for their missions.[17] As America wallowed through its economic crisis in early 2009, supporters of spaceflight found that their job included more than persuading skeptical lawmakers in Congress. Even newly elected President Barack Obama seemed to be uncertain on whether the destiny of NASA and the United States would include people on the planet Mars.

During a brief meeting with reporters from regional newspapers, President Obama took a single question on the condition of the nation's space program. He declared NASA to be "adrift" and in need of a refined mission.[18] That statement rankled the agency's usually stoic associate administrator for space operations. In March 2009, Bill Gerstenmaier had just attended the launch of Space Shuttle *Discovery* on its mission to deliver the final set of solar electricity-generating panels to the International Space Station. The launch had impressed even longtime Shuttle managers. The vehicle blasted off at sunset, and the rays of the waning day hit *Discovery*'s smoky exhaust trail at such an angle that it created bands of white, gold, peach, gray, and black. Not since the launch of *Atlantis* to deliver the International Space Station *Destiny* lab in 2001 had launch observers enjoyed such a dramatic view. *Discovery*'s liftoff also had skies so clear that observers on the ground could see the vehicle drop its solid rocket boosters. NASA officials estimated *Discovery* was still visible from Florida as the spacecraft passed New York and New Jersey on its trip to orbit. During the traditional postlaunch press conference, the subject eventually turned to the president's comment about NASA being adrift. Gerstenmaier was asked to respond. "The agency is on track to finish the space station," he responded. "And we need to keep doing the job of letting Washington and the people know what it is that we do here."[19] That battle will likely drag on, assuming the Orion capsule makes its first docking flights to the International Space Station, and NASA plans its return to the Moon as a possible precursor to the trip to Mars that the agency and space enthusiasts have been patiently waiting decades to pursue. Even diehard supporters admit it's a voyage that may never come. In the meantime, kids will no doubt continue to dream of being astronauts, and astronauts will continue to dream of going to Mars.

epilogue

When the planet Venus is particularly bright in the nighttime sky or appears close to the Moon, it may rate a mention in the local newspaper. Events involving Mars draw crowds. Going there in a spacecraft will be as dangerous as trailblazing the Old West, but looking through a telescope just takes patience.

In August 2003, thousands of schoolchildren and their hapless parents gathered outside the Orlando Science Center on a school night. It wasn't for the latest release in the Harry Potter book series, but rather to see Mars. The facility in Loch Haven Park near downtown routinely staged events to try to attract young patrons, like IMAX movies or the "Our Body" exhibit, which featured human cadavers in various poses. This evening, all the Science Center had to do was handle the overwhelming crush of people.[1]

Earth and the red planet would be closer together in their orbits around the Sun than they had been in 60,000 years, and Mars madness had ensued. The line at the Science Center snaked around the building as people waited for their turn to have a look at Mars through the facility's telescope, with its ten-inch lens. The wait in line at the Crosby Observatory, with its silver dome, was comfortable, until you got inside. The atmosphere then turned claustrophobic as the crush of people made their way up a narrow steel spiral staircase to the telescope. That's where the treasured eyepiece allowed each visitor to take a brief peep.

The view was good, but not great. Ironically, after waiting in line for hours for the highly advertised observatory telescope, it was the amateurs who saved the day. Local astronomy buffs had toted their fourteen-inch telescopes and set up shop on the walkway surrounding

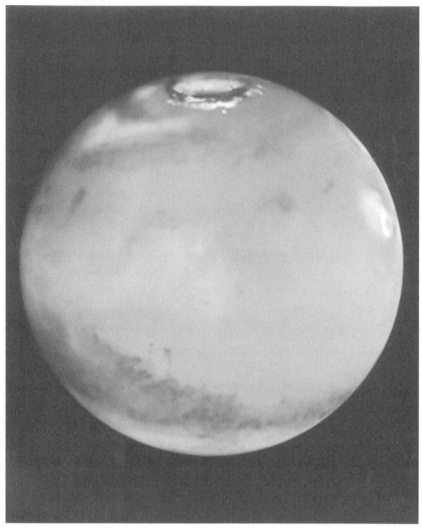

This photograph of Mars was taken by the Hubble Space Telescope in 2003, during the closest pass between Earth and Mars in 60,000 years. The most prominent feature is the southern polar cap at the bottom. Courtesy of NASA.

the observatory dome. Here, people could look as long as they liked into the clear nighttime sky.[2]

And Mars blazed.

The most obvious feature that was visible on the burnt orange–colored globe was the southern polar cap forming a small, rough white spot on the bottom. With a little educated coaxing from the astronomers, more landmarks could be made out. A long darkish splotch across the center of the planet, separating the northern lowlands and southern highlands, was Syrtis Major. It had been first spotted by Dutch astronomer Christiaan Huygens in the late 1600s. He later became the namesake of a tiny probe dropped into the gaseous cauldron that made up the atmosphere of one of Saturn's moons. The *Huygens* lander hitched a ride to the planet aboard NASA's *Cassini* spacecraft. The circular patch to the lower left on Mars was Hellas Planitia. Scientists believed this feature of the planet was an ancient impact crater, and possibly the home of an ocean in the distant past.[3]

Punctuating the excitement of the evening was something too small to be seen even with the Hubble Space Telescope. Two months prior to the close encounter between Mars and Earth, NASA had launched its two exploration rovers. The two robotic vehicles were well on their way to the red planet, but too tiny for the visitors to the Science Center to spot, no matter how hard they squinted. The $400 million spacecraft each went by the dull, technical designations of MER-A and MER-B— that is, until two months before the crowds gathered at the Orlando Science Center.

NASA held a contest among schoolchildren to give the rovers more interesting names. In June 2003, nine-year-old Sofi Collis's suggestions of *Spirit* and *Opportunity* were picked as the winners. Sofi had grown up in an orphanage in Siberia before being adopted by a couple from Scottsdale, Arizona. Officials at the space agency couldn't help but note how the third grader represented not one but two space-faring nations.[4] The event would give NASA the chance to think about something other than the pain of the Space Shuttle *Columbia* accident that had occurred in January of that year.

Supporters of the space program gathered at Walt Disney World's Epcot Center in April 2004 for the opening of an exhibit on the Mars rovers. The ceremony was held outside the theme park's new "Mission:

Space" attraction, which included a huge glossy model of the planet Mars out front in the ride's courtyard. A quote from NASA's chief administrator Sean O'Keefe was now emblazoned across it. "We're back, and we're on Mars," it read. Sofi Collis was there as well, in a kid-size pair of blue NASA flight coveralls. The youngster stole the show with the quick snap of her head while being introduced and the perky responses to questions from the reporters in attendance.[5] But while NASA basked in the glow of the rovers, and fans of Mars strained for a look at the red planet, America's manned spaceflight program was slowly outgrowing its mission. It was a process that had taken decades.

Between the time of the first Space Shuttle launch in 1981 and the *Challenger* disaster that killed seven astronauts in 1986, NASA conducted its experimental missions with the Shuttle. These early trips confirmed the flightworthiness of the vehicle, the ability of the astronauts to do spacewalks, and the Shuttle's capacity to launch commercial satellites. The space agency had yet to launch its own network of tracking and data relay satellites to beam pictures and computer data to Earth, so the video sent down from the spacecraft had the same grainy and otherworldly quality that people saw from the Apollo Moon landings in the 1960s and 1970s.

After *Challenger* exploded, NASA's recovery included redesigning the Shuttle's solid rocket boosters. Once flights resumed, the agency moved into a period of more routine trips to orbit, referred to as the "lark" missions. The Shuttle had a list of interplanetary probes to launch, like *Magellan* to the planet Venus and *Galileo* to Jupiter, both in 1989. There were also science missions to study the atmosphere and to practice construction on the nation's yet-to-be-built space station. The agency finished launching its network of communication satellites, so photos of stars and galaxies from the Hubble Space Telescope came down in vibrant color and clarity. In 1995, the United States and Russia began to collaborate on the joint docking flights between the Space Shuttle and the Russian space station *Mir*.

Then the program took its next turn with the building of the International Space Station in 1998. At that point, the Shuttle finally lived up to its reputation as America's "space truck" by carrying up the U.S.-built crew modules; sections of the outpost's long, spinelike outer framework; and the feathery solar panels that would provide the bulk of the electricity for the orbiting complex. The finished space station would be the size of a football field.

Finally, the *Columbia* accident in 2003 would spell the eventual end of the Shuttle program and the birth of the Orion space capsule, with the goal of sending people to the Moon and perhaps to Mars. Like a college graduate who moved back into mom and dad's house, NASA was outgrowing its current situation and needed to grow up and move on to something else. That may, or may not, include Mars. However, changes are already apparent at the space agency.

Take astronaut Mike Fincke, for example. While his colleagues in NASA's astronaut corps competed for one of the final seats on the agency's last Shuttle missions, Fincke had been in space twice. But he didn't ride on an American spacecraft on either mission. Both trips were aboard Russian-built Soyuz capsules, and Fincke is the only astronaut to hold that distinction. "The big difference, as I understand," Fincke recalled of riding the Russia craft as opposed to the Shuttle, "was that the Soyuz was a lot smoother. However, the landings are more, shall we say, impactful."[6] That's because instead of landing with wings or plopping down in the ocean, Soyuz capsules use parachutes and booster rockets to come down on solid ground in the steppes of Kazakhstan. Fincke's second mission aboard the International Space Station also demonstrated the hazards of spaceflight in orbit around Earth, let alone the dangers of a voyage to Mars. In early 2009, the outpost had three encounters with "space junk." That's the term used for leftover hardware from rocket launches that whizzes around the planet in its own orbit.

NASA tries to track this trash from the ground, but it's not always easy. Imagine a baseball outfielder watching a pop fly hit from home plate. If the ball goes into the Sun, the fielder knows the ball is up there somewhere, and it's moving in his direction. But with the glare of sunlight, it's impossible to keep an eye on the ball the whole time. That's NASA's problem. Mission managers try to compute the path space junk takes around Earth, but it's an inexact science.

During Mike Fincke's mission on the International Space Station, one piece of debris came so close to the station that he and his two crewmates were ordered into their Soyuz space capsule in case they needed to use it as a lifeboat if the outpost were to be hit. The approaching threat that day was a piece off of a small booster rocket for a satellite.

It was no bigger than a person's thumb.[7]

Crews venturing to Mars will likely face greater dangers than this. The list includes radiation exposure, the hazards of vital equipment failures, and medical emergencies, not to mention the death-defying descent to the planet's surface and the return blastoff to Earth. After decades of relative comfort of flights in low Earth orbit or on brief trips to the Moon, NASA and the American public will enter the "big leagues" of space travel, assuming a manned Mars mission is ever attempted.

Like the mostly anonymous pioneers who tamed the West, these astronauts will likely face casualties along the way. It remains to be seen if the United States will be ready to lose astronauts in a space program that, up to now, has been regarded as a novelty or a stunt. That may be the biggest test as proponents of a Mars mission prepare for trailblazing Mars, NASA's next giant leap.

acknowledgments

The only thing more challenging than publishing your first book is publishing the second one. I would like to express my appreciation to these people whose support helped to make *Trailblazing Mars* possible.

Robert E. Witt, president of the University of Alabama (UA)
Judy Bonner, executive vice president and provost, UA
Loy Singleton, dean, UA's College of Communication and Information Sciences

Elizabeth Brock, director, Alabama Public Radio, UA

June Malone and Dave Drachlis, Marshall Space Flight Center

Lisa Malone, Bill Johnson, Manny Virata, Allard Beutel, Candrea Thomas, Laurel Lichtenberger, Margaret Persinger, Kay Grinter, and Lesley Garner, Kennedy Space Center

Guy Webster, NASA's Jet Propulsion Laboratory

Gayle Frere, Johnson Space Center

Carleton Bailie

notes

Chapter 1. Mariner Sets Sail

1. Media History Project, University of Minnesota, http://www.mediahistory. umn.edu/timeline/1960–1969.html.

2. Stamatios Krimigis, interview by author, July 2008.

3. Louis Friedman, "Remembering James Van Allen," *Planetary News*, August 2006, http://www.planetary.org/news/2006/0810_Remembering_James_Van_Allen.html.

4. Ibid.

5. Robert Roy Britt, "Shoot from the Hip: A History of Rocketry," July 24, 2001, Space.com.

6. Kevin Nolan, *Mars: A Cosmic Stepping Stone* (New York: Copernicus Books, 2008).

7. "Mariner 1 and 2: Quick Look," Jet Propulsion Laboratory, http://msl.jpl.nasa.gov/QuickLooks/mariner12QL.html.

8. Ibid.

9. David Portree, "Humans to Mars," *Monographs in Space History* 21 (2001): 13-23.

10. Krimigis, interview.

11. Ibid.

12. John Casani, interview by author, September 2008.

13. Tara Gray, "Alan Shepard," 40th Anniversary of the Mercury 7, 1999, http://www.hq.nasa.gov/office/pao/History/40thmerc7/shepard.htm.

14. J. Randy Taraborelli, *Jackie, Ethel, Joan: Women of Camelot* (New York: Warner Books, 2000).

15. "Mariner 3 & 4: Quick Look," http://msl.jpl.nasa.gov/QuickLooks/mariner34QL.html.

16. Arvydas Kliore, interview by author, July 2008.

17. Ibid.

18. Ibid.

19. Ibid.

20. Marc Leslie Kutner, *Astronomy: A Physical Perspective* (Cambridge: Cambridge University Press, 2003).

21. Mary Decker, "Watery Surprises on Mars," *Fate Magazine*, June 2001.

22. Frank Salisbury, "Martian Biology: Accumulating Evidence Favors the Theory of Life on Mars, but There May Be Surprises," *Science*, April 1962.

23. Bruce Murray, interview by author, September 2008.

24. Kliore, interview.

25. "On Mars: Exploration of the Red Planet," NASA History, http://history.nasa.gov/SP-4212/ch3.html.

26. *Time*, March 8, 1963.

27. Ibid., July 23, 1965.

Chapter 2. The Space Race to Mars

1. Tara Gray, "Gordon Cooper," *40th Anniversary of the Mercury 7*, http://history.nasa.gov/40thmerc7/cooper.htm.

2. "The History of Jodrell Bank," http://www.jodrellbank.manchester.ac.uk/history/.

3. Ibid.

4. Andrew Lepage, interview by author, January 2009.

5. Ibid.

6. "Zond-2," National Space Science Data Center, http://nssdc.gsfc.nasa.gov/nmc/spacecraftDisplay.do?id=1964–078C.

7. Lepage, interview.

8. Larry Klaes, interview by author, January 2009.

9. Ibid.

10. Ian O'Neill, "The Mars Curse: Why Have So Many Missions Failed?" *Universe Today*, March 22, 2008, http://www.universetoday.com/2008/03/22/the-mars-curse-why-have-so-many-missions-failed/.

Chapter 3. Reheating the Leftovers of Apollo

1. Paul A. Cantor, *Gilligan Unbound: Popular Culture in the Age of Globalization* (Lantham, Md.: Rowman and Littlefield, 2003).

2. Hamish Lindsey, *Tracking Apollo to the Moon* (New York: Springer Books, 2001).

3. David Portree, "Humans to Mars," NASA, http://www.nss.org/settlement/mars/2001-HumansToMars-FiftyYearsOfMissionPlanning.pdf.

4. Barton C. Hacker and James M. Grimwood, *On the Shoulders of Titans: A History of Project Gemini*, Chapter 12, http://history.nasa.gov/SP-4203/ch12–1.htm.

5. Alan Bean, interview by author, September 2008.

6. David Baker, *The Rocket: The History and Development of Rocket and Missile Technology* (London: New Cavendish, 1978).

7. James Dewar, *To the End of the Solar System* (Lexington: University of Kentucky Press, 2004).

8. John F. Kennedy, "Special Message to Congress," May 25, 1961, http://www.jfklibrary.org/Historical+Resources/Archives/Reference+Desk/Speeches/JFK/003POF03NationalNeeds05251961.htm.

9. Ibid.

10. Harold Finger, interview by author, June 2008.

11. Ibid.

12. Bean, interview.

13. Walter Cunningham, interview by author, September 2008.

14. David R. Williams, "Apollo 18 through 20: The Cancelled Missions," December 2003, http://nssdc.gsfc.nasa.gov/planetary/lunar/apollo_18_20.html.

15. Cunningham, interview.

16. Bean, interview.

17. Ibid.

18. Lisa Tidwell, "Skylab: Stepping Stone to Today's ISS," *Space Center Roundup* (2003), 6-8.

19. Ibid.

20. Ibid.

21. Ibid.

22. Phillip Baker, *The Story of Manned Space Stations* (New York: Springer, 2007), 40.

23. Ibid.

24. Warren Leary, "NASA Sets Next Mars Mission," *International Herald Tribune*, August 9, 2005.

25. Michael Lemonick, "Mars Reconsidered," *Time*, December 12, 1999, http://www.time.com/time/magazine/article/0,9171,992882,00.html.

26. Daniel Goldin, interview by author, January 1999.

Chapter 4. Viking, NASA's "Gold Bug"

1. "NASA's Huntress on Mars," CNN, August 1996.

2. Peter Smith, interview by author, September 2008.

3. Harold Klein, "The Viking Mission Search for Life on Mars," NASA Ames Research Center, http://mars.spherix.com/spie2/Reprint75.htm.

4. Casani, interview.

5. Ibid.

6. Murray, interview.

7. Angelo Guastaferro, interview by author, September 2008.

8. Stephen Wilkinson, "Mach 1: Assaulting the Barrier," Air & Space Smithsonian, December 1990, http://www.airspacemag.com/history-of-flight/Mach_1.html.

9. "Exploring NASA's Roots: The History of the Langley Research Center," http://www.nasa.gov/centers/langley/news/factsheets/LaRC_History.html.

10. Guastaferro, interview.

11. Ibid.

12. Murray, interview.

13. Ibid.

14. Ibid.

15. "Mars Facts," Cornell University, http://athena.cornell.edu/mars_facts/past_missions_70s.html.

16. Guastaferro, interview.

17. Ibid.

18. Murray, interview.

19. NASA 10th anniversary observance of Viking, July 1986, National Academy of Science, Washington, D.C.

20. STS-51L press kit, 1986, history.nasa.gov/sts51lpresskit.pdf.

21. "Report of the Presidential Commission on Space Shuttle Challenger Accident," 1986, http://science.ksc.nasa.gov/shuttle/missions/51-l/docs/rogers-commission/table-of-contents.html.

22. Ibid.

23. Ibid.

24. NASA 10th anniversary observance of Viking.

25. *Time*, October 20, 1980.

26. STS-63 press kit, 1995, http://science.ksc.nasa.gov/shuttle/missions/sts-63/sts-63-press-kit.txt.

27. Carl Sagan, press interviews, National Academy of Science, Washington, D.C., July 1986.

28. Ibid.

29. Ibid.

30. Ibid.

31. Gilbert Levin, interview by author, August 1986.

32. Ibid.

33. Ibid.

34. Ibid.

35. Ibid.

36. Gerald Soffen, interview by author, August 1986.

37. Levin, interview.

38. Ibid.

39. "Pioneer 10, 11, Quick Look," http://msl.jpl.nasa.gov/QuickLooks/pioneer10QL.html.

40. "Voyager Forges a New Frontier," NASA, http://www.nasa.gov/missions/deepspace/voyager.html.

41. Murray, interview.

Chapter 5. The Twenty-Year Gap

1. NASA Discovery Workshop Summary, November 1992, http://openlibrary.org/b/OL17682155M/NASA_Discovery_Program_Workshop.

2. Smith, interview.

3. STS-34 press kit, 1989, http://science.ksc.nasa.gov/shuttle/missions/sts-34/sts-34-press-kit.txt.

4. "Probe Sees Titan's Methane Clouds," BBC News, April 2004.

5. "Cassini Races toward Saturn Orbit," CNN, June 2004.

6. Fraser Cain, "Huygens Celebrates a Year on Titan," *Universe Today*, January 2006, http://www.universetoday.com/2006/01/14/huygens-celebrates-a-year-on-titan/.

7. Ibid.

8. Smith, interview.

9. Ibid.

10. Ibid.

11. Ibid.

12. Daniel Britt, interview by author, September 2008.

13. R. Rieder, H. Wanke, and T. Economou, "An Alpha Magnetic Spectrometer for Mars 96 and Mars Pathfinder," abstract delivered to American Astronomical Society, September 1996.

14. Smith, interview.

15. Britt, interview.

16. Ibid.

17. Ibid.

18. Ibid.

19. Smith, interview.

20. Ibid.

21. Ibid.

22. Ibid.

23. Ibid.

24. Britt, interview.

25. Smith, interview.

26. Anthony Young, *Lunar and Planetary Rovers: The Wheels of Apollo and the Quest for Mars* (New York: Springer, 2007).

27. Britt, interview.

28. Ibid.

29. Ibid.

30. Smith, interview.

31. Oliver Morton, "Lessons from the Frontier," *Newsweek*, December 13, 1999, http://www.newsweek.com/id/90505.

32. Britt, interview.

33. "Earth to Mars Climate Orbiter: Are We There Yet?" *Science Daily*, September 13, 1999.

34. Robin Lloyd, "Metric Mishap Caused Loss of NASA Orbiter," CNN, September 30, 1999.

35. Smith, interview.

36. Thomas Young, testimony before the House Science Committee, April 2000, http://www.spaceref.com/news/viewpr.html?pid=1444.

37. Ibid.

38. Ibid.

39. Jonathan Amos, "Probe Ends Historic Mars Mission," BBC News, November 10, 2008.

40. Smith, interview.

Chapter 6. The Shuttle's Long Good-bye

1. Marcia Dunn, "NASA Estimates 3,000 to 4,000 Job Losses," Associated Press, June 23, 2008.

2. Jefferson Morris, "NASA Workforce Projections Previewed," *Aviation Week*, February 23, 2008.

3. Ibid.

4. Bill Nelson, NASA biography, http://www.jsc.nasa.gov/Bios/htmlbios/nelson-b.html.

5. Marcia Dunn, "NASA: Fewer Workers to Lose Their Jobs," Associated Press, June 23, 2008.

6. "NASA May Have to Depend on Russia for Space Travel," *Manufacturing & Technology News*, January 24, 2008, 14.

7. Richard Dunham, "Evolving Obama Now Supports $2 Billion More for NASA," *Houston Chronicle*, August 18, 2008.

8. Ibid.

9. John Torres, "McCain's Straight Talk Express Rolls into Brevard," *Florida Today*, August 18, 2008.

10. Howard McCurdy, interview by author, August 2008.

11. John Schwartz, "The Fight Over NASA's Future," *New York Times*, December 29, 2008.

12. John Logsdon, "Ten Presidents and NASA," NASA's 50th, nasa.gov.

13. Meeting of the Presidential Commission on Human Spaceflight, Cocoa Beach Hilton, July 30, 2009.

14. Ibid.

15. Ross Thierny, interview by author, July 2009.

16. "Astronaut Video Satirizes NASA Bureaucracy," *Morning Edition*, National Public Radio, February 9, 2009.

17. Thierny, interview.

18. Ibid.

19. Norman Augustine press briefing, Cocoa Beach Hilton, July 2009.

20. NASA Columbia Crew Survivability Report, December 2008, http://www.nasa.gov/pdf/298870main_SP-2008-565.pdf.

21. STS-114 press kit, August 2005, http://www.nasa.gov/pdf/112301main_114_pk_july05.pdf.

22. Warren Leary, "NASA Chief Affirms Stand on Canceling Hubble Mission," *New York Times*, January 29 2004.

23. Scott Altman, interview by author, February 2005.

24. John Grunsfeld, interview by author, October 2008.

25. Jeff Hecht, "The Testing Error That Led to Hubble Mirror Fiasco," *New Scientist*, August 18, 1990, http://www.newscientist.com/article/mg12717301.000-the-testing-error-that-led-to-hubble-mirror-fiasco.html.

26. Ibid.

27. W. Patrick McCray, *Giant Telescopes* (Cambridge, Mass.: Harvard University Press, 2004).

28. Shirish Date, "Latest Photos From Space Improve Hubble's Image," *Orlando Sentinel*, January 14, 1994.

29. Grunsfeld, interview.

30. Ibid.

31. Michael Foale, interview by author, June 2008.

32. Grunsfeld, interview.

33. Ibid.

34. STS-125 press kit, 2008, http://www.nasa.gov/pdf/331922main_sts125_presskit_050609.pdf.

35. Grunsfeld, interview.

36. Ibid.

37. Ibid.

38. Ibid.

Chapter 7. Pioneers Past and Present

1. Jerry Ross, interview by author, September 2008.

2. Williams, "Apollo 18 through 20."

3. Bettyann Holzman, *Almost Heaven: The Story of Women in Space* (New York: Basic Books, 2003).

4. STS-61B press kit, 1985, www.jsc.nasa.gov/history/shuttle_pk/.../Flight_023_STS-61B_Press_Kit.pdf.

5. Ross, interview.

6. Ibid.

7. STS-27 press kit, 1988, http://www.shuttlepresskit.com/STS-27/STS27.pdf.

8. Ross, interview.

9. "NASA's Successful Gamma Ray Observatory Mission Comes to an End," *Science Daily*, May 27, 2000.

10. Ross, interview.

11. Ibid.

12. "Russia Prepares to Launch First Part of International Space Station," CNN, October 2, 1998.

13. Susan Erler, "Texas Auction House Says It Has the Wooden Gun Used in C. P. Jail Escape," *Northwest Indiana Times*, October 8, 2009.

14. Ross, interview.

15. Ibid.

16. John Lane, interview by author, September 2008.

17. Ibid.

18. Ibid.

19. Lucy Sheriff, "NASA Reveals Manned Mars Mission Plans," *Register*, November 29, 2007.

20. NASA Lunar Architecture press conference, December 2006.

21. Lane, interview.

Chapter 8. The Moon, One Baby Step

1. "Apollo Chronicles: The Mysterious Smell of Moondust," http://www.nasa.gov/exploration/home/30jan_smellofmoondust.html.

2. Arthur C. Clarke, KSC press conference, August 1994.

3. Jacques Cousteau, KSC press conference, November 1994.

4. "Arthur C. Clarke Dies at 90," March 18, 2008, Space.com.

5. STS-68 press kit, http://www.shuttlepresskit.com/sts-68/index.htm.

6. STS-68 mission overview, NASA, http://science.ksc.nasa.gov/shuttle/missions/sts-68/mission-sts-68.html

7. Ibid.

8. Clarke, press conference.

9. Robert Zimmerman, *Leaving Earth* (Washington, D.C.: Joseph Henry Press, 2003).

10. Wilson DeSilva, "Children of Apollo," *Cosmos*, December 2006, http://www.cosmosmagazine.com/features/print/1163/children-apollo.

11. "Hilton Studies the Feasibility of Building a Space Hotel," *Guardian*, September 27, 1999.

12. Paul Spudis, testimony before the U.S. House Science and Technology Committee, April 2004, http://www.spaceref.com/news/viewsr.html?pid=12407.

13. NASA Lunar Architecture press conference.

14. Ibid.

15. Lewis Page, "John Glenn Blasts Moonbase to Mars NASA Roadmap," *Register*, August 1, 2008, *http://www.theregister.co.uk/2008/08/01/john_glenn_says_moonbase_ploy_questionable/.*

16. "Moon Base Must Precede Mars Mission," September 30, 2008, Space.com.

17. Marc Cohen, interview by author, October 2008.

18. Ibid.

19. Maggie McKee, "One Rover Gets Stuck In, Another Chills Out," *New Scientist*, June 1, 2006, http://www.newscientist.com/article/dn9253-one-mars-rover-gets-stuck-in-the-other-chills-out.html.

20. Brian Wilcox, interview by author, October 2008.

21. Ibid.

22. Ibid.

23. Ibid.

24. Ibid.

25. Jonathan Fildes, "Inflatable Space Module Puffs Up," BBC News, July 14, 2006.

26. Ibid.

27. STS-77 press kit, http://science.ksc.nasa.gov/shuttle/missions/sts-77/mission-sts- 77.html.

28. David Cadogan, interview by author, October 2008.

29. Ibid.

30. Ibid.

31. Ibid.

32. Jason Bryenton, interview by author, November 2008.

33. Ibid.

34. Ibid.

35. Ibid.

36. Geoffrey Yoder, interview by author, November 2008.

37. "The Rendezvous That Was Almost Missed," NASA Facts On Line, December 1992, http://oea.larc.nasa.gov/PAIS/Rendezvous.html.

Chapter 9. The New "Right Stuff"

1. "Astronaut Selection," in *This New Ocean*, Johnson Space Center, http://history.nasa.gov/SP-4201/ch6–8.htm.

2. Ibid.

3. Ibid.

4. Scott Horowitz, interview by author, November 2008.

5. Ibid.

6. Ibid.

7. STS-113 Terminal Countdown Demonstration Test press conference, October 27, 2002.

8. STS-97 press kit, 2000, http://www.shuttlepresskit.com/sts-97/.

9. "Crew Cleared in Soyuz Landing," Associated Press, May 26, 2003.

10. STS-126 press kit, November 2008, http://www.nasa.gov/pdf/287211main_sts126_press_kit2.pdf.

11. NASA-TV, STS-126 prelaunch coverage, November 2008.

12. Ibid.

13. STS-126, postlaunch press conference, November 2008.

14. Muriel Draaisma, "The ISS Renovation Project," Canadian Broadcasting Company, November 13, 2008.

15. STS-126 press kit.

16. Ibid.

17. Seth Borenstein, "Spacewalkers Add to Orbiting Junk," Associated Press, September 14, 2006.

18. Marcia Dunn, "Finally, Urine Recycler Passes Astronaut Test," Associated Press, November 25, 2008.

19. Tariq Malik, "Astronaut Perfects Space Coffee with Zero-G Cup," November 24, 2008, Space.com.

20. Horowitz, interview.

21. Ibid.

22. Geoffrey Yoder, interview by author, December 2008.

23. Lawrence Palinkas, interview by author, October 2008.

24. Ibid.

25. Horowitz, interview.

26. Ibid.

27. Palinkas, interview.

28. Andrew Allen, interview by author, July 1999.

29. STS-46 press kit, http://www.shuttlepresskit.com/sts-46/index.htm.

30. Allen, interview.

31. Ibid.

Chapter 10. Lessons from Biosphere 2

1. Baker, *Story of Manned Space Stations*, 88.

2. STS-63 press kit.

3. Jane Poynter, interview by author, November 2008.

4. Rene Gutel, "Biospherians Remember Life in a Self-Contained World," VOA news, 2006, http://origin.www.voanews.com/english/archive/2006-05/2006-05-02-voa21.cfm?moddate=2006-05-03.

5. Poynter, interview.

6. Ibid.

7. Ibid.

8. Ibid.

9. Ibid.

10. Julian Dibbell, "The Fast Supper," *New York Magazine*, October 22, 2006, http://nymag.com/news/features/23169/.

11. Poynter, interview.

12. Ibid.

13. Ibid.

14. Ibid.

15. Ibid.

16. Archive for Center for Research on Controlled Ecological Life Support Systems at Purdue University 1990–1995, Purdue University.

17. Cary Mitchell, interview by author, November 2008.

18. Roger Lewin, "Living in a Bubble," *New Scientist*, April 2002, http://www.newscientist.com/article/mg13418152.900-focus-living-in-a-bubble—biosphere-2-was-billed-as-the-model-for-human-colonies-in-space-but-critics-argue-that-the-outside-world-has-been-duped.html.

19. Poynter, interview.

20. Ibid.

21. Raymond Cronise, "Self-Organized Critical in Closed Ecosystems: Carbon Dioxide Fluctuations in Biosphere 2," NASA Marshall Spaceflight Center, 1995.

22. Ibid.

23. Poynter, interview.

24. Mitchell, interview.

25. Poynter, interview.

Chapter 11. Plowing the "Back 40" on Mars

1. "The Decision to Go to the Moon," NASA Office of History, http://history.nasa.gov/moondec.html.

2. K. R. Sridhar, interview by author, December 2008.

3. Sean Alfano, "Space Shuttle Launch Postponed," CBS News, September 8, 2006.

4. Rex Hall and David Shayler, *Soyuz: A Universal Spacecraft* (New York: Springer, 2003).

5. Horowitz, interview.

6. Harold McGee and Daniel Patterson, "Talk Dirt to Me," *New York Times*, May 6, 2007.

7. "Mars Soil Could Support Life," BBC News, June 27, 2008.

8. Mitchell, interview.

9. Ibid.

10. Ibid.

11. Ibid.

12. Maurice Averner, "The NASA Celss Program," NASA headquarters, http://ntrs.nasa.gov/archive/nasa/casi.ntrs.nasa.gov/19910004535_1991004535.pdf.

13. Horowitz, interview.

14. "New Amundsen-Scott Station," National Science Foundation, http://www.nsf.gov/news/special_reports/livingsouthpole/station_new.jsp.

15. Gioia Massa, interview by author, November 2008.

16. Robert Block, "Endeavour Lifts Off on Extreme Design Mission," *Los Angeles Times*, November 15, 2008.

17. Mitchell, interview.

18. Sridhar, interview.

19. Ibid.

20. Mitchell, interview.

21. Sridhar, interview.

Chapter 12. Should NASA Go It Alone?

1. F. Robert Van der Linden, ed., *Best of the National Air and Space Museum* (Washington, D.C.: Smithsonian Institution, 2006).

2. "A New Proposal," NASA History, http://history.nasa.gov/SP-4209/ch5–2. htm.

3. Ibid.

4. Carl E. Lewis, "Soviet Cooperation in Space: A Case Study," http://www.global security.org/space/library/report/1989/LCE.htm.

5. Gus W. Weiss, "The Farewell Dossier: Duping the Soviets," Central Intelligence Agency, 2006, https://www.cia.gov/library/center-for-the-study-of-intelligence/kent-csi/pdf/v39i5a14p.pdf.

6. Brian Harvey, *Russia in Space: The Failed Frontier* (New York: Springer Praxis, 2001).

7. "Oral History Transcript," Arnold Frutkin, interview by Rebecca Wright, Johnson Space Center, 2002. www.jsc.nasa.gov/history/oral_histories/NASA. . ./FrutkinAW_3-8-02.pdf.

8. STS-48 press kit

9. Reightler, interview by author, February 2009.

10. Ibid.

11. Kathy Laurini, interview by author, February 2009.

12. STS-58 press kit

13. STS-55 press kit

14. STS-122 press kit

15. "Columbus," Dornier GmbH, March 1998.

16. Hans Schlegal, interview by author, October 2006.

17. Laurini, interview

18. "Ariane 5 and *Jules Verne* Arrive at Launch Pad," European Space Agency, March 2008, http://www.esa.int/esaCP/SEMIWWK26DF_index_0.html.

19. "Multi Purpose Logistics Module," Marshall Space Center, March 2001, http://mplm.msfc.nasa.gov/.

20. Laurini, interview.

21. Tariq Malik, "From Earth to the Station: Europe's First Space Cargo Ship," March 5, 2008, Space.com.

22. Laurini, interview.

23. Reightler, interview.

24. Ibid.

25. Ibid.

26. STS-60 press kit, http://science.ksc.nasa.gov/shuttle/missions/sts-60/sts-60-press-kit.txt.

27. Sergei Krikalev, NASA biography, Johnson Space Center, http://www.jsc.nasa.gov/Bios/htmlbios/krikalev.html.

28. Reightler, interview.

29. Ibid.

30. Ibid.

31. Krikalev, NASA biography.

32. Reightler, interview.

33. Yuri Karash, "Liftoff! Dennis Tito, First Space Tourist Safely in Orbit," April 28, 2001, Space.com.

34. STS-107 crew walkout, Kennedy Space Center, Florida, January 16, 2003.

35. STS-60 prelaunch breakfast, NASA-TV, February 3, 1994.

36. Bruce Melnick, interview by author, June 2005.

37. Reightler, interview.

38. Ibid.

Chapter 13. Getting There, Living There

1. Alan Shepard, interview by author, August 1994.

2. Sam Beddingfield, interview by author, June 2005.

3. "Apollo 10 Objectives," NASA History, http://history.nasa.gov/SP-4029/Apollo_10b_Objectives.htm.

4. W. David Woods, *How Apollo Flew to the Moon* (New York: Springer Verlag, 2007).

5. David Bromwich, "William Safire: Wars Made Out of Words," Huffington Post, October 1, 2009.

6. Keith Campbell, "NASA Unveils New Thinking on Human Missions to Mars," *Engineering News*, December 14, 2007.

7. Augustine Commission meeting, Cocoa Beach Hilton, July 2009.

8. "Human Exploration of Mars: The Reference Mission of NASA's Mars Exploration Study Team," July 1997. ftp://nssdcftp.gsfc.nasa.gov/. . ./mars. . ./mars_ref_mission_sp6107.pdf.

9. Ibid.

10. Jim Wilson, "Bringing Life to Mars," *Popular Mechanics*, November 1998, http://www.popularmechanics.com/science/air_space/1282466.html.

11. Campbell, "NASA Unveils New Thinking."

12. Ibid.

13. Horowitz, interview.

14. Ibid.

15. Yoder, interview, December 2008.

16. Robert Zubrin, interview by author, January 2009.

17. Ibid.

18. Ibid.

19. "NASA's Dirty Little Secret: Moon Dust," *Science Daily*, September 29, 2008, http://www.sciencedaily.com/releases/2008/09/080924191552.htm.

20. John Mustard, interview by author, January 2009.

21. Ibid.

22. Ibid.

23. Ibid.

24. NASA Lunar Architecture press conference.

Chapter 14. Why Go at All?

1. *The Space Shuttle Decision*, Chapter 4, "Winter of Discontent," http://history. nasa.gov/SP-4221/ch4.htm.

2. Cunningham, interview.

3. Stephen Weinberg, interview by author, March 2009.

4. "NASA Science Missions Getting Cut," *Universe Today*, March 12, 2006, http:// www.universetoday.com/2006/03/12/the-nasa-science-missions-getting-cut/.

5. Weinberg, interview.

6. Gregg Easterbrook, "Beam Me Out of This Deathtrap, Scotty," *Washington Monthly*, April 1980, http://www.washingtonmonthly.com/features/2001/8004. easterbrook-fulltext.html.

7. Vincent Canby, "Moonraker," *New York Times*, June 29, 1979.

8. Gregg Easterbrook, interview by author, March 2009.

9. Ibid.

10. "NASA Authorization Act of 2008," *Congressional Record*, May 16, 2008 110th Congress, http://legislative.nasa.gov/PL%20110-422.pdf.

11. "50th Anniversary of the Space Race," NASA, headquarters.http://www.nasa. gov/externalflash/SpaceAge/.

12. Ailish O'Hora, Brendon Keenan, and Michael Brennan, "Shares Plunge Following the St. Patrick's Day Massacre," *Independent*, March 18, 2008.

13. "Frank Calls Mission to Mars 'Flying Pork,'" http://www.c-spanarchives.org/ congress/?q=node/77531&id=7417768.

14. Barney Frank, interview by author, March 2009.

15. Zubrin, interview.

16. David Weldon, interview by author, October 2008.

17. Ibid.

18. Mark Mathews, "Obama Calls NASA an Agency Adrift," *Orlando Sentinel*, March 11, 2009.

19. STS-119 postlaunch press conference, March 15, 2009.

Epilogue

1. "Mars to Get Closer to Earth Ever in Recorded History," November 8, 2002, Space.com.

2. Orlando Science Center, Mars viewing, August 1, 2003.

3. François Forget, François Costard, and Phillipe Lognonne, *Mars: Story of Another World* (New York: Springer, 2006).

4. "Mars Rovers Named *Spirit* and *Opportunity*," June 8, 2003, Space.com.

5. "NASA Mars Success Honored at Disney Day of Discovery," *Orlando Business Journal*, April 2004, http://orlando.bizjournals.com/orlando/stories/2004/04/05/ daily20.html.

6. Bob Edwards, "NASA Update," *Bob Edwards Weekend*, March 21, 2008.

7. "Space Station Crew Takes Temporary Shelter in Soyuz," Associated Press, March 12, 2009.

suggested reading

Barbree, Jay. *"Live from Cape Canaveral": Covering the Space Program, from Sputnik to Today*. New York: Collins, 2008.

Barnaby Faherty, William. *Florida's Space Coast: The Impact of NASA on the Sunshine State*. Gainesville: University Press of Florida, 2002.

Cernan, Eugene, and Donald Davis. *The Last Man on the Moon: Astronaut Eugene Cernan and America's Race in Space*. New York: St. Martin's Press, 1999.

Cunningham, Walter. *The All American Boys*. Anaheim, Calif.: I Books, 2003.

Hansen, James. *First Man: The Life of Neil A. Armstrong*. New York: Simon and Schuster, 2005.

Kranz, Gene. *Failure Is Not an Option: Mission Control from Mercury to Apollo 13 and Beyond*. New York: Simon and Schuster, 2000.

Lipartito, David, and Orville R. Butler. *A History of the Kennedy Space Center*. Gainesville: University Press of Florida, 2007.

Mullane, Michael. *Riding Rockets: The Outrageous Tales of a Space Shuttle Astronaut*. New York: Scribner, 2007.

Murray, Charles, and Catherine Bly Cox. *Apollo: The Race to the Moon*. New York: Simon and Schuster, 1989.

Poynter, Jane. *The Human Experiment: Two Years and Twenty Minutes Inside Biosphere 2*. New York: Basic Books, 2006.

Shelton, William. *Man's Conquest of Space*. Washington, D.C.: National Geographic Society, 1972.

Thompson, Milton O. *At the Edge of Space: The X-15 Flight Program*. Washington, D.C.: Smithsonian Institution Press, 1992.

index

Pat Duggins is news director at Alabama Public Radio. He has spent over twenty years providing national stories on the space program, including the development of NASA's next generation spacecraft at the Marshall Space Flight Center in Huntsville. He covered more than one hundred space shuttle missions for National Public Radio, starting with the 1986 *Challenger* accident. Following the 2003 loss of *Columbia*, Pat provided three hours of live coverage on NPR's Weekend Edition with Scott Simon. He is the author of *Final Countdown: NASA and the End of the Space Shuttle Program*.